OTHER BOOKS BY RICHARD MORRIS

Light
The End of the World
The Fate of the Universe
Evolution and Human Nature
The Word and Beyond (coauthor)

DISMANTLING
The Nature

THE UNIVERSE

of Scientific Discovery

Richard Morris

A TOUCHSTONE BOOK
Published by Simon & Schuster, Inc.
NEW YORK

First Touchstone Edition, 1984

Published by Simon & Schuster, Inc.
Simon & Schuster Building
Rockefeller Center
1230 Avenue of the Americas
New York, New York 10020

TOUCHSTONE and colophon are registered trademarks
of Simon & Schuster, Inc.

Designed by Eve Kirch

Manufactured in the United States of America

10 9 8 7 6 5 4 3 2 1
10 9 8 7 6 5 4 3 2 1 Pbk.

Library of Congress Cataloging in Publication Data

Morris, Richard, date.
    Dismantling the universe.
    Bibliography: p.
    Includes index.
    1. Science—Methodology. 2. Creative ability in
science. I. Title.
Q175.M869   1983   501   83-11324

ISBN 0-671-45239-8
ISBN 0-671-52818-1 Pbk.

CONTENTS

PREFACE

This is a book that began to change almost as soon as I started to write it. I had originally intended to write only about creative imagination as an element in scientific discovery. But as I began to research the topics that I planned to cover, I realized that matters were not so simple as I had imagined.

I found that I was asking myself questions that I had not previously considered. For example, certain scientific theories are frequently said to be "beautiful." But what does beauty have to do with truth? How can a theory that is motivated by a quest for beauty turn out to be an accurate description of nature?

Another question had to do with experimental verification. I began to realize that theories are often accepted long before empirical proof becomes available. For example, Copernicus published his hypothesis concerning a moving earth and a sun-centered solar system in 1543. Yet it was not until 1851 that physicists were able to devise an experiment which proved, beyond all possible doubt, that the earth really did rotate on its axis. And yet the Copernican hypothesis was accepted long before this experiment was performed. By 1727—the year of Newton's death —any scientist who refused to believe in a heliocentric solar system would have been considered a crackpot.

I had hardly begun to consider these questions when others began to occur to me. One of the first to come to mind was the following: Exactly how does one distinguish between a plausible scientific theory which has not been verified and a crackpot idea which is not supported by any empirical evidence either? How can physicists be so certain about the reality of subatomic particles that have yet to be discovered, and yet unhesitatingly dismiss the idea that extraterrestrial beings might be visiting the earth in UFOs? How can astronomers be so sure that the existence of black holes (for which there is only circumstantial evidence at present) will eventually be verified, and yet be so vehement in their dismissal of the idea that the earth could have been created in 4004 B.C.?

These are just some of the questions that I found myself considering as I went along. I am not sure that the answers I have found are the only possible ones. But I do think that my attempts to answer them illuminate my main topics: the nature of scientific discovery, its logic and illogic, and its relation to other kinds of human creativity.

Since I was trained as a physicist, I have found it natural to take many of my examples from the fields of physics and astronomy. But I have not hesitated to draw topics from a number of other scientific fields as well. The subject of this book, after all, is not physics but rather scientific discovery in general. I have found that examples taken from such disciplines as geology and chemistry often provided the best illustrations of the points I wanted to make.

I have used the metric system throughout. Since this system of measurement is not yet completely familiar to many American readers, I have frequently translated quantities such as kilometers into miles and put the latter in parentheses. I have also made frequent use of exponential notation to express large numbers. It is much more convenient to say "10^{26}" than it is to write the numeral 1 and twenty-six zeros.

I have used the masculine generic pronoun fairly consistently. Constructions such as "he or she" sound awkward to me. Since there does not yet seem to be any new usage that is generally agreed upon, I have stuck to the old one on the theory that it is still standard. I am not sure that I even agree that "he or she" is less sexist. When that meaning is intended, it can generally be inferred from the context with little difficulty.

Finally, I would like to thank Dr. William T. Scott for reading and commenting upon parts of the manuscript. His knowledge of certain of the topics discussed in this book is unequaled, and his advice was extremely valuable.

Chapter 1

THE BREAKDOWN
OF COMMON SENSE

A BLACK HOLE is a collapsed relic of a dead star which has a gravitational field so intense that nothing, not even light, can escape from it. The mass of a black hole is concentrated in a small volume in its center called the *singularity*. The singularity is surrounded by a surface called the *event horizon*. The size of the event horizon is determined by the amount of mass that the black hole contains. The diameter of the event horizon is approximately 6 kilometers for every solar mass. Thus a black hole that had ten times the mass of our sun would be about 6o kilometers across.

The gravitational forces in a black hole become very large, perhaps even infinite, at the singularity. Even at the event horizon, they become immensely strong. They are so strong that time itself seems to come to a stop. There exist mathematical equations which describe this situation precisely. It is a little more difficult to talk about it in ordinary language. However, it would not be too inaccurate to say that a black hole is not only a hole in space, it is also a hole in time.

Perhaps an example will make this bizarre-sounding statement a little clearer. Let us imagine that a manned space probe is falling into a black hole. This is something that may never happen. If the time ever comes when we can actually journey to black holes, it is not likely that anyone will venture into one willingly. Anyone who did would be committing suicide. Nevertheless, we can certainly perform a "thought experiment" and imagine that someone has.

Let us imagine also that the probe has maintained television communication with a mother ship which has remained far enough away from the black hole to experience no significant gravitational effects. As the probe nears the event horizon, it will appear to observers on the mother ship that events on the probe are happening more and more slowly. Clocks will slow down, equipment will begin to operate in slow motion, and human speech will slow to an incomprehensible drawl.

But the people on the probe will observe none of these things. From their point of view, everything will remain perfectly normal. The probe will continue to function in its ordinary manner. Within a relatively short time, they will have passed the dark event horizon and will be able to see the singularity itself. When this happens, they will know that death is not far away. If their ship has so far escaped destruction, the intense gravity to which it is being subjected will quickly rip it apart.

The observers on the mother ship will never see this happen. Since, from their viewpoint, time on the space probe is slowing down, the probe will seem to approach the black hole at an ever-more-leisurely pace. The probe will approach nearer and nearer to the event horizon, but it will never cross it. Eventually, time on the probe will freeze, and it will remain poised just outside the black hole's surface.

From the standpoint of an outside observer, nothing can ever penetrate a black hole's event horizon. Whatever approaches it is destined to remain there for as long as the universe lasts. And yet from the standpoint of an observer who is falling into a black hole, nothing of the sort happens. There is nothing to prevent him from entering the black hole's interior in a matter of minutes in his time frame. Time on the ship is so compressed that billions of years in the outside universe seem to be only minutes or seconds to an observer who is approaching the event horizon of the black hole. When he enters the hole an infinite length of time will have passed on the outside.

One feels almost compelled to object that such things could not possibly happen. The idea that the same time span could seem to be a few minutes to one observer and an eternity to another appears fundamentally illogical. It is a gross violation of common sense.

Yet according to Einstein's general theory of relativity, this is exactly what does happen. Furthermore, we have every reason to believe that the predictions of the theory are correct. During the 1960s and 1970s, the general theory of relativity was confirmed by numerous experimental tests of high accuracy. In every case, its predictions were confirmed. Although the theory might fail under extreme conditions, such as those which exist near the singularity, there is every reason to believe that it can describe the conditions at the event horizon of a black hole with near-perfect accuracy.

In the view of most contemporary physicists, the objection that general relativity seems to violate common sense is no criticism at all. Common sense, after all, is a product of our experience with ordinary objects that are moving about in an ordinary manner on the surface of the earth. There is no reason to believe that objects which are subjected to very intense gravitational fields should behave in the same manner as those which remain in weak ones. For example, we have every reason to expect that a football will behave in a reasonably normal manner when it is thrown from one point to another on the earth's surface. If a quarterback threw a pass and the ball remained frozen above the 40-yard line for all eternity, we would have good reason to be surprised. On the other hand, there is really no good reason for believing that our commonsense ideas about the behavior of footballs should remain applicable under the extreme conditions that exist in the vicinity of a black hole.

For that matter, we should not expect that our commonsense ideas should be valid under any extreme conditions. There is no reason to think that objects which move at velocities approaching the speed of light should behave like footballs, or automobiles, or other objects which move about at everyday velocities. Neither is there any good reason for thinking that our commonsense expectations should apply in the world of the very large or in that of the very small. We should expect that the behavior of subatomic particles would be bizarre. We should expect that scientific ideas concerning the nature of the universe would seem strange to someone who is confronting them for the first time.

It is not always easy to give up our commonsense outlook. The difficulty does not stem so much from the need to discard old concepts as it does from the apparently paradoxical nature of new scientific ideas. For example, it is not hard to accept the notion that a particle such as an electron cannot be expected to behave the way a football does. It is more difficult to figure out why an electron should sometimes appear to be a particle and sometimes look like a bundle of waves. It is even harder to understand how an electron can act like a particle and yet lack such characteristics as a definite position and a definite velocity.

But perhaps it would be best if we postponed a discussion of electrons and other subatomic particles until later. Since I have begun by talking about the strange behavior of time in the vicinity of a black hole, it might be best to take a closer look at the theories which changed our conceptions of space and time before we go on to other subjects. So suppose we go back to the year 1905, the year in which the breakdown of common sense first became glaringly apparent.

It can be said that every new scientific theory involves the overthrow of common sense. Science perpetually seeks to penetrate the appearances of the everyday world in order to discover a more fundamental reality. As it departs from the familiar, it commonly discovers that things are not quite what they seem to be. The discovery that the earth moves around the sun overthrew a commonsense idea. So did the discovery that the stars are other suns, not tiny lights in the sky. But in 1905, the breakdown suddenly accelerated. In that year, the German physicist Albert Einstein discovered that nature was even stranger than anyone had imagined it could be. In that year, he dismantled the universe that physicists had been constructing over a period of several centuries.

Einstein published three revolutionary papers in 1905. All appeared in the German scientific journal *Annalen der Physik*. Each one had a profound effect on the course of modern physics. One of these papers, which had the rather prosaic title "On the Movement of Small Particles Suspended in a Stationary Liquid Demanded by the Molecular-Kinetic Theory of Heat," established once and for all time the reality of atoms. In 1905, this was a topic about which some physicists still entertained doubts. The second paper, "On a Heuristic Point of View about the Creation and Conversion of Light," showed that light, which had previously been thought to be a wave phenomenon, sometimes behaved as though made up of particles. The third paper, "On the Electrodynamics of Moving Bodies," has become the most famous. In it, Einstein propounded his special theory of relativity.

Textbooks and popular works alike often begin their accounts of special relativity with a discussion of the Michelson-Morley experiment, which was first performed by the American scientists Albert Michelson and Edward Morley in 1887. The implication is that the special theory was Einstein's response to Michelson and Morley's puzzling experimental result.

There is some justification for this. The result of the Michelson-Morley experiment was surprising to the physicists of the day. Einstein's theory did successfully explain why such a strange result had been obtained. However, it is not completely certain that Einstein knew of the Michelson-Morley experiment in 1905. In 1950, Einstein stated that he had not heard of the experiment when he wrote his paper on relativity. Had he known of it, he asserted, he certainly would have mentioned it in the paper. Two years later, in 1952, he contradicted himself; he said that he had been aware at the time of the Michelson-Morley result.

Einstein was then more than seventy years old, and his memory may have been playing tricks on him. However, it really makes little difference whether he first became acquainted with the Michelson-Morley result before or after 1905; on more than one occasion, he made it clear that the experiment had had, at most, a negligible effect on his thinking. The connection between Michelson and Morley and Einstein seems to have been invented by scholars who wanted to make the development of modern physics seem a little more logical than it really was.

They should not be judged too harshly for this. The Michelson-Morley result was puzzling to the physicists of the day. And a discussion of it can show why a theory of relativity was needed. So perhaps it would be best if I commented upon the experiment before going on to a discussion of the ideas in Einstein's paper.

In 1887, it had been known for some time that light was a wave phenomenon. Physicists made the commonsense assumption that waves could not exist without something to vibrate in. It was known that sound consisted of vibrations of the air, just as ocean waves were made possible by seawater movement. Scientists reasoned, by analogy, that light was caused by vibrations in a hypothetical substance called the *ether*.

The conclusion was drawn that the ether must permeate all space. Otherwise, light could not travel to the earth from the stars or from the sun. But this seemingly natural conclusion led to grave difficulties. If the ether existed, it could not be a form of ordinary matter. Furthermore, it had to have some very odd properties. It had to be rigid enough to transmit the extremely rapid light vibrations, and it had to be so rarefied as not to impede the motions of the planets.

In spite of all the problems that the concept of the ether presented, few scientists doubted that the substance existed. Therefore it seemed natural to conduct an experiment that would detect it. This was the experiment that Michelson and Morley performed in 1887.

The two scientists realized that if the earth was moving through the ether, then a ray of light that was traveling in the same direction as the earth should have a lower velocity relative to us than one that was traveling in the opposite direction. After all, there should be an ether "wind" blowing past the earth. Light should travel more slowly when it had to fight its way upstream than it did when it had the wind at its back.

To Michelson's and Morley's surprise, no significant ether wind could be detected. They were certain that the fault did not lie with the experiment. It was accurate enough to detect an ether wind that had a velocity much less than that of the earth.*

This result did not cause scientists to give up on the ether concept. It did not occur to them that the idea should be abandoned. Some suggested that the earth somehow dragged the ether along with it. Others proposed that perhaps motion through the ether caused solid bodies to contract in the direction of their motion. If the contraction had just the right magnitude, they pointed out, then the change in the length of Michelson and Morley's experimental instruments would account for the null result.

If Einstein had pondered the result of the Michelson-Morley experiment, he undoubtedly would have been led to the special theory of relativity. Indeed, shortly after Einstein's theory was published, it was realized that special relativity showed why Michelson and Morley had obtained such a strange result.

However, Einstein did not begin his paper with a discussion of this experiment. He began by discussing an old idea called *relativity*. He found that if he combined the assumption that the principle of relativity was valid with the postulate that the speed of light did not depend upon the velocity of the source that emitted it, some very surprising conclusions could be drawn.

The idea that all motion was relative was not original with Einstein. It had been noted long before 1905 that the laws of mechanics, the branch of physics dealing with the motion of bodies, depended only on the relative motion of the objects considered. If a weight was dropped from the mast of a moving ship, it did not make any difference how fast the ship was moving across the ocean; the path of the weight relative to the ship would be the same whether the ship was moving or stationary.

This principle is called *Galilean relativity*, after the Italian

---

* Strictly speaking, the Michelson-Morley result was not entirely negative. They did obtain a small positive result. However, the result was much smaller than they expected; it was almost certainly the result of minor experimental inaccuracies.

scientist Galileo, who had proposed it around the beginning of the seventeenth century. Although no one doubted that the principle was true, it took an Einstein to see its implications.

Einstein noted that some of the laws of physics, in particular those dealing with electricity and magnetism, did not embody the principle of relativity. So Einstein began his paper with a discussion of the motion of magnets and electrical conductors. Since *electrodynamics* was the name that had been given to the theory which described electricity and magnetism, Einstein called his paper "The Electrodynamics of Moving Bodies."

When Einstein wrote his paper on relativity, it had long been known that a moving magnet creates an electric field. There is nothing very mysterious about this; it is the principle upon which electrical generators are based. A generator is simply a device which uses rotating magnets to set up an electrical current in a wire. The moving magnets generate an electrical field, and the electrical field induces a current.

The laws of electrodynamics also imply that, if a magnet is held stationary and a wire near it is moved, the same electrical current should be set up. This is precisely what one should expect to happen. After all, it shouldn't make any difference whether it is the wire or the magnet that is moved. The only thing that should make any difference is their relative motion.

In both cases, the predictions of the theory were correct. Theory said that a current should be set up, and a current was observed. Nevertheless, there was something very peculiar about the theory. It did not describe the two cases in the same way. When the magnet was moved, an electric field was set up. It was this field which created the current in the wire. When the magnet was motionless, on the other hand, no such field was created, even though the same current was induced.

In other words, the theory of electrodynamics distinguished between a moving magnet and a moving wire. In one case there was an electric field; in the other, there was not. In electrodynamics, unlike mechanics, there was such a thing as absolute motion.

Einstein could not accept this. Consequently, he made the assumption that the principle of relativity should apply to electrodynamics also. Then he attempted to see how the laws of electrodynamics could be modified so that they would contain only the idea of relative motion.

Einstein found that if he wanted to modify the laws of electrodynamics, it was necessary to make an assumption from which mathematical consequences could be deduced. The idea that the velocity of light was constant gave the desired result. It did not make any difference whether the source was moving or not, and it didn't make any difference what state of motion the observer was in. The velocity of light was always 300,000 kilometers (186,000 miles) per second.

This was quite an audacious assumption. After all, the velocities of ordinary objects do not exhibit this kind of constancy. If a jet plane is traveling at a speed of 1,000 kilometers per hour, and if it fires a rocket that has a velocity of 800, the rocket will streak toward its target at 1,800 kilometers per hour; the two speeds must be added. The assumption that light did not behave in this manner seemed a violation of common sense.

It was this assumption which explained the result of the Michelson-Morley experiment. If it was correct, then it was only natural that Michelson and Morley should have found that light had the same velocity when it traveled in the direction of the earth's motion as it did when it traveled in any other direction. Einstein's assumption implied, in addition, that the notion of an ether was superfluous. If light did not travel relative to a stationary medium, then it was not reasonable to assume that any such medium existed. Light, it seemed, could propagate without any ether in which to "vibrate."

If all Einstein had done was show that the concept of an ether could be discarded, that would have constituted a significant achievement. However, he did much more. He discovered that the principle of relativity and the assumption of the constancy of the speed of light had numerous other implications. Not only did the special theory of relativity clear up a problem concerning the behavior of magnets, it also had far-reaching implications concerning the behavior of bodies in motion.

One mathematical consequence of Einstein's theory was that the physical dimensions of bodies depended upon the relative motion of objects and observers. For example, suppose that a space vehicle is traveling away from me at a velocity that is 90 percent of the speed of light. It will seem to me that the ship has contracted to about one-half its original length. Now, this is really not surprising. This is the same contraction that was predicted by the physicists who had been looking for a way to account for the failure of the Michelson-Morley experiment.

However, there is a significant difference between the outlook of those physicists and that of Einstein. They believed that the contraction was "real"—that it was the same for any observer. Einstein, on the other hand, saw it as something that was only relative. According to Einstein's theory, the occupants of the spaceship will not observe any change in their vessel. From their standpoint, it is I who will have contracted to one-half my original length (or height, if the ship is moving in a vertical direction). Since all motion is relative, they have as much justification for claiming that it is I who have shrunk as I have for maintaining that they have contracted.

The special theory also has important implications concerning the nature of time. It implies, first, that no two events that are separated in space can be said to be simultaneous in an absolute sense. If one observer sees event A and event B happen at the same time, then a second observer may see A happen first if he is in a different state of motion. A third observer, furthermore, may see B as happening before A. And since the viewpoint of one observer is as good as the viewpoint of any other, all three interpretations are equally valid. Thus the notion of "simultaneity" as applied to spatially separated events is a relative concept. Although we can talk about two events that both happen at, say, 1:38 P.M. in a given place, it is meaningless to say that an event on a distant star or planet "happens at the same time" as one on earth. Different observers will never agree as to whether they do or not.

The special theory also predicts a *time dilation*, or slowing down of time, that is analogous to the contraction in length. When I watch a spaceship travel away from me at 90 percent of the velocity of light, everything that happens on the ship will appear to be happening only half as fast. Similarly, the occupants of the ship will see me perform all my actions in slow motion.

One of the most important implications of special relativity has to do with the mass of a moving body. There exists a *relativistic mass increase*. If a spaceship accelerates to a velocity approaching that of light, I will think that its mass is increasing. Again, the effect works in reverse: the crew of the ship will think that it is I who am getting heavier.

The existence of a relativistic mass increase has two important consequences. First, it implies that no material object can ever reach the velocity of light. The faster it goes, the greater its mass will be. But the increased mass causes greater resistance to further acceleration. If an object were to attain the velocity of light, its mass would become infinite. An infinite amount of energy would be required to bring this about. The situation from the viewpoint of the ship's crew is even worse. From the standpoint of an observer on the ship, every object in the outside universe would be required to accelerate to infinite mass. This would require infinite energy for each one of them.

It is futile to attempt to find loopholes in this argument. It is perfectly straightforward and convincing. Furthermore, there exist other arguments which indicate that the speed of light represents a barrier that cannot be crossed. If an object could somehow attain the speed of light, it would disintegrate instantly. Atoms are held together by electromagnetic forces. Like light (which is a form of electromagnetic radiation), these forces are propagated at a velocity of 300,000 kilometers per second. If an object could attain a velocity greater than this, the forces that held its constitutent atoms together would be left behind. Even if there were a way to get around the infinite-mass problem, a faster-than-light spaceship would evaporate into a cloud of subatomic particles.

Finally, the existence of a "light barrier" is confirmed by experiment. When particles are accelerated in particle accelerators, they approach the velocity of light, but never attain it. If enough energy is given to an electron or a proton, it may be accelerated to a speed that is 90 percent or even 99 percent of the velocity of light. A figure of 100 percent is never reached.

The special theory of relativity does not, however, entirely rule out the possibility that there might be particles which travel faster than light. It is conceivable that such particles might exist. But if they did, they would encounter the same barrier from the other side. They could never slow down to velocities which are less than that of light.

Such hypothetical particles are called *tachyons*. There is really no good reason for thinking that they exist. There is no experimental evidence for them. Furthermore, special relativity only suggests them as an interesting theoretical possibility; it does not say that they must be real.

In fact, there are good reasons for doubting the existence of tachyons. If such particles did exist, they could travel backward in time. This implies that it would be possible to send signals from the future into the past. The only way to avoid the paradoxes that result from such a possibility would be to assume that tachyons cannot interact with ordinary matter. However, the assumption that tachyons exist but cannot be detected renders the question of their reality rather metaphysical.

Such considerations have not stopped physicists from performing experiments designed to detect tachyons on the assumption that perhaps they can interact with matter. However, the results have so far been negative, or ambiguous at best. Consequently, it would be best to remain skeptical about their existence until such time as there exists some evidence to the contrary.

The existence of a relativistic mass increase has one other important implication. When an object is accelerated to a high velocity, an expenditure of energy is required. This suggests a connection between the energy expenditure and the mass increase. In fact, if the two quantities are equated, one obtains the famous equation $E = mc^2$. This suggests that mass and energy are equivalent. If one can be transformed into the other in this case, there is no reason why they cannot be transformed under other conditions also.

Since Einstein derived his equation in 1905, this conclusion has been confirmed over and over again. Mass is changed into energy every time a radioactive atom decays. It is the transformation of mass into energy that accounts for the heat and light emitted by the sun, and for the energy released in the explosion of a nuclear bomb. The reverse process—transformation of energy into mass —is observed somewhat less frequently in nature. However, it has been observed in the laboratory on numerous occasions.

The special theory successfully accounts for phenomena that occur when velocities are constant; but it does not describe how things look to an observer who is accelerating. If an object is made to go faster and faster, what additional effects will this produce? Einstein realized that if he could answer this question also, he might be able to understand the nature of gravity. After all, the outstanding characteristic of a gravitational field is that it causes objects to accelerate. If an object is dropped from a height, it does not fall at a constant velocity. On the contrary, it will have a velocity of 9.8 meters (32 feet) per second after one second has passed. After two seconds, the velocity will have doubled. At the end of ten seconds (provided the effects of air resistance are minimal), it will be ten times as great.

Einstein worked on a theory that would describe such phenomena—his general theory of relativity—for years. The general theory was finally published in 1916. Although it successfully described how things should appear when seen from the viewpoint of an accelerated observer, its most significant implications had to do with the nature of gravity, and the effects that a gravitational field should have on objects subjected to its influence.

It is the general theory of relativity which predicts that gravity will cause time to slow down in the vicinity of a black hole. Actually, the theory implies that this will happen in any gravitational field, even in a relatively weak one such as that of the earth. However, since the gravitational forces created by a black hole are much more intense, the effects will be more dramatic.

The slowing of time in the relatively weak field of the earth has been confirmed by experiment. It has been found that accurate atomic clocks speed up to a small degree when they are sent aloft in rockets. Since the earth's gravity becomes weaker at higher altitudes, this is just what one would expect to happen if the theory is correct. Other experiments have involved measurements of the bending of light rays and radio waves by the sun and the detection of radar echoes from various planets in the solar system. Various different kinds of astronomical objects have been found to exhibit behavior predicted by general relativity. One of the most dramatic confirmations of the theory came in 1979 when a *gravitational lens* was discovered. In that year, astronomers discovered that gravitational forces exerted by a distant galaxy had distorted the light coming from a distant starlike object to such a great extent that its image was split when it was photographed through telescopes; it appeared as two objects instead of one.

According to the general theory of relativity, gravity can distort both space and time. It produces the dramatic time dilation that would be observed near a black hole, and it is responsible for what is known as "curved space."

I must confess that I have always felt a little uncomfortable with this term. "Curved space" is an accurate enough concept provided that one states precisely what one means by it. But its use in books for the general reader is sometimes misleading. Space, after all, is not an object which can be warped in the same manner as a material body. As Einstein himself pointed out, "space is not a thing."

The term "curved space" (which, incidentally, Einstein did not use in his 1916 paper) is the result of attempts to describe the content of Einstein's rather complicated mathematical equations in ordinary English. As is often the case, something is lost in the translation. When we say that space is "curved," what we really mean is that light rays travel through it in a way that cannot be described by the ordinary Euclidean geometry that is taught to high school students. The geometry of Einsteinian space is somewhat different.

A simple example should make this idea somewhat clearer. In Euclidean geometry, the angles of a triangle always add up to 180 degrees. It doesn't make any difference what the triangle looks like, or how big the individual angles are; their sum is always the same. Now, it so happens that mathematicians have long known that other kinds of geometry are possible. There exist geometrical systems with properties unlike those of Euclidean geometry. One of these properties is that the sum of the angles of a triangle is *not* 180 degrees.

Physicists paid little attention to these *non-Euclidean* geometries when they were discovered. It was believed that only Euclidean geometry had any application to the real world; the other varieties were thought to be nothing more than mathematical abstractions.

In 1916, Einstein showed that this assumption was mistaken. In the presence of a gravitational field, the sum of the angles of a triangle is not 180 degrees. For example, suppose that three observers are communicating with each other by means of light beams. The light rays that travel between them will define a triangle. If each observer measures the angle between the rays at his position, and if the figures are added up, the result will be either more than 180 degrees, or less. Space is Euclidean only when gravitational fields are absent.

Gravitational fields not only change the geometry of space; they affect the "geometry" of time as well. Space and time become intermixed. The forces produced by gravitating bodies cannot distort one without also altering the other. The mathematical equations that describe this situation are much more complicated than those of Newtonian physics, where the space and time dimensions can be kept separate.

When gravitational fields are very strong, the changes in the geometry of space and time can be striking. They are so striking, in fact, that commonsense notions cease to be a reliable guide. For example, it is this change in geometry which makes it impossible for anything to escape from the interior of a black hole. If a ray of light is directed toward a black hole, there are no problems; there are numerous geometrical paths by which it can cross the event horizon. However, there exist no paths which lead from the interior of a black hole to the outside universe. If a spaceship managed to enter a black hole, it could presumably still beam a ray of light in any direction. However, whatever direction was chosen, the light could never reach the event horizon from the inside. As far as any observer inside a black hole is concerned, the event horizon no longer exists. It is possible to think of the interior of a black hole as a region sealed off from the rest of the universe. It is spatially sealed off, and sealed off in the time dimension as well.

The Einsteinian universe was a strange one. At least it appeared strange to physicists accustomed to the "commonsense" views that were embodied in nineteenth-century physics. However, there was really nothing paradoxical about Einstein's conclusions. Physicists who studied his theories in detail soon recognized that Einstein had introduced his new ideas about the nature of time and space not because he was in love with the bizarre, but because he wanted to weld physics into a harmonious whole. He had propounded his theories of relativity because he had felt that the old physics was not as logical or consistent as it should be.

Although there was some initial opposition to Einstein's ideas, physicists soon realized that relativity must be taken seriously. The inner logic that could be perceived in the theories was perhaps even more convincing than experimental confirmations. In fact, if Einstein's theories had not been logically convincing, scientists would probably not have taken them seriously enough to perform the experiments that were needed to substantiate them. After all, scientists do not waste valuable time testing ideas that they think to be crackpot. Few of them desire to submit papers to scientific journals which state that they had failed to confirm a theory which probably hadn't been true in the first place.

A factor that may have contributed to the relatively rapid acceptance of Einstein's ideas was the fact that during the years in which relativity was formulated, a second revolution in physics was in progress. By the time the special theory was published in 1905, scientists were already beginning to realize that they might be forced to accept strange new ideas if they were to have any hope of understanding the nature of atoms. Some of them had already begun to understand that nature was not always what it seemed to be to the naive observer.

This second revolution had begun during the closing years of the nineteenth century, when the German physicist Max Planck began to try to find the solution to one of the outstanding unsolved problems of physics, that of the nature of *blackbody radiation.*

Blackbody radiation is radiation that is emitted from a perfectly black object when it is heated. It is similar to the radiation that is emitted when a stone or a piece of iron is placed in a fire. An object heated in this manner will begin to glow red when it reaches a temperature of about 600 degrees Celsius. When it is made hotter, it will turn orange, then yellow and finally white.

When physicists investigate a phenomenon, they always attempt to study the simplest cases first. It is best not to introduce complications at the beginning; after all, the more complicated situations can always be looked into later. Thus, when they began to study the properties of glowing objects, they preferred not to study stones, or pieces of iron, but objects that were perfectly dark when they were cool.

It so happens that there is one small problem. No such objects exist. Even soot reflects some of the light that falls on it. However, a little ingenuity allows this obstacle to be overcome. If a small hole is punched in a box, then the hole will be perfectly black. Any light that passes in through the hole will be absorbed by the interior walls of the box; none will be reflected back out.

If the box is made of a substance that will not burn when it is heated—a metal, for example—blackbody radiation can be easily studied. As the box is heated, it will begin to glow. Light will be emitted from the outside walls of the box, and from the inside as well. Some of the light given off from the inside walls will make its way out through the hole. The radiation passing through it will be identical to that which would be given off by an object that is perfectly black.

When this experiment was performed, the results did not correspond to the predictions of existing theories. In fact, accepted theory gave results that were absurd. It said that an infinite amount of energy should be given off at short wavelengths.

It was obvious that this was impossible. A finite amount of heat could clearly not produce infinite quantities of radiation. Furthermore, observations indicated that the amount of energy radiated did not always go up as wavelengths got shorter. An object that is red hot, for example, gives off most of its energy in the form of red light. Although the wavelengths of blue light are shorter than those of red, blue light is hardly given off at all.

When Planck began to wrestle with the problem, there had been previous attempts to solve it. However, all of these had failed. It seemed that whatever assumptions one added to the existing theory, incorrect results were always obtained.

So Planck decided to try a different approach. Instead of beginning with a set of reasonable-sounding assumptions that could be incorporated into the theory, Planck set theoretical questions aside. He began by looking for a mathematical formula that would describe blackbody radiation. Once a formula was found, he tried to see what assumptions had to be made if the formula was to be valid.

Planck found, to his surprise, that the formula implied that light could be emitted only in discrete packets of energy. That is, a blackbody could emit one unit of energy, or four, or 165,007,-713,982. But it could not give off one and a half, or six and seven-tenths, or 132,873 and one-eighth.

This was a very surprising result. According to the wave theory of light, it should have been possible for energy to be radiated in any quantity whatsoever. If the properties of waves were what physicists believed them to be, it was absurd to suggest that light was given off in little bundles. Nevertheless, there seemed to be no way to avoid this conclusion. So, in 1900, Planck announced his *quantum theory* of light. Light, he said, was emitted in packets, or *quanta*. This was the only way that blackbody radiation could be explained.

This was a revolutionary idea. However, Planck proved to be a very reluctant revolutionary. He had hypothesized the existence of quanta because he could see no way to avoid such a conclusion. However, he could not help viewing the idea with distaste. Speaking of the formulation of the quantum theory many years later, Planck described it as "an act of desperation." After his theory was published, Planck struggled to find a way to undo his own work. He was a conservative physicist who hoped that a way could be found to avoid this odd assumption, and to reinstate the classical wave description of the nature of light.

On the other hand, Einstein, who had a somewhat more audacious temperament, possessed no such reservations. In 1905, the same year that he propounded his special theory of relativity, he suggested an explanation for Planck's mystifying result. Perhaps, Einstein said, light was made up of particles. If one made this assumption, he pointed out, then Planck's quantum theory was easy to understand. The reason that only whole numbers of light quanta were emitted became obvious. Either a particle was given off or it wasn't. Particles were not emitted in halves, or in thirds, or in any other fraction.

Furthermore, Einstein pointed out, the assumption that light had a particle nature provided a solution for some of the other problems that had been bedeviling physicists. The hypothesis not only cast light on the nature of blackbody radiation; it also explained such phenomena as photoluminescence, and the emission of electrons from metals that were illuminated with ultraviolet light.

Einstein was aware that evidence for the wave nature of light had been accumulating for more than two centuries. He knew very well that this evidence was so conclusive that it could not be dismissed. And yet he went ahead and proposed that light had a particle nature nevertheless.

To some of the scientists of the day, it must have seemed that Einstein had lost his mind. If he was right, the only conclusion one could draw was that light was made up of waves and particles at the same time. However, "wave" and "particle" were two mutually exclusive categories. Light could not be both at the same time, they objected.

As one might expect, Einstein's theory was not widely accepted. There was widespread opposition to it, and as late as 1917, the American physicist Robert Millikan commented in his book *The Electron* that the theory had proved to be "so untenable that Einstein himself, I believe, no longer holds to it."

But Millikan was wrong. Einstein had not repudiated the theory. Six years later, his faith in the idea was vindicated. In 1923, the American physicist Arthur Holly Compton performed some experiments with X rays which demonstrated that Einstein had been right after all.

Like light, X rays are a form of electromagnetic radiation. The only difference between the two is that the wavelengths of X rays are much smaller. An experiment which showed that X rays were composed of quanta (or *photons*, as physicists call them today) would imply that light had a particle nature too. One could not reasonably assume that two types of radiation which were fundamentally alike had a different composition.

Compton observed that when X rays were passed through paraffin, some of them were scattered in various different directions. This was not unexpected. It stood to reason that some of the X rays should bounce off the paraffin's constituent atoms. This would cause them to be reflected at some angle to the original X-ray beam.

However, when this happened, something unexpected was observed. Compton found that the wavelengths of the scattered X rays were changed. There was no way that the wave theory could explain this. X rays that were reflected from atoms should no more change their wavelengths than light that was reflected from a mirrror.

When Compton tried applying the quantum theory to the phenomenon, he found that the wavelength change could easily be understood. If X rays were composed of particles, those which collided with atoms should give up some of their energy in the collisions. According to the theory that Planck had discovered in 1900, wavelength and energy were related. The shorter the wavelength, the greater was the amount of energy that each photon possessed. If the quantum theory was correct, a change in wavelength and a change in energy were really the same thing.

Although some scientists believed that Compton had demonstrated that light (or X rays) was made of particles, others continued to reject the concept of the dual nature of light as too paradoxical. The British physicist Sir William Bragg sarcastically remarked that there seemed to be no recourse but to believe in the wave theory on Monday, Wednesday and Friday, and to use the particle theory on Tuesday, Thursday and Saturday. Bragg made no mention of Sunday because the six days between Monday and Saturday made up the British academic week. It didn't take long before some anonymous wit had filled the gap, however. "And on Sunday," he added, "we pray for guidance."

The scientists who had been brought up on classical nineteenth-century theories continued to hope that this odd paradox would eventually be resolved. But to many of the members of the younger generation of physicists, it did not seem to be so much of a paradox at all. Rather than spend their time trying to imagine how light could be waves and particles at the same time, these younger scientists set to work developing quantum theory even further.

One of these young physicists was a Frenchman named Louis de Broglie. In 1924, De Broglie submitted a doctoral dissertation to his professors at the École Polytechnique in Paris in which it was suggested that if light could be made up of particles, then perhaps particles could be viewed as waves. Such subatomic particles as electrons, De Broglie pointed out, might exhibit a wave character.

De Broglie's examiners did not know what to make of the theory. Uncertain as to whether De Broglie should be passed or failed, one of the professors showed the dissertation to Einstein, who replied that in his view, De Broglie had made a fruitful suggestion that might turn out to be an important discovery. Shortly afterward, Einstein published a paper in which he called attention to De Broglie's results.

Many physicists were skeptical. One of them went so far as to refer to De Broglie's theory as *la Comédie Française*. A few years later, he may have regretted that he had laughed at it, for De Broglie's idea was soon confirmed by experiment. A phenomenon called *electron diffraction* was observed by the physicist George Thomson in England, and by the young physicists Clinton Davisson and Lester Germer in the United States. Thomson and the team of American experimenters independently observed that interference patterns could be obtained when electrons were made to bounce off a crystal onto a photographic plate. When the image formed by the electrons was examined, it was found to be made up of light and dark fringes that were very similar to those which were formed when a crystal was bombarded with X rays.

In the case of X rays, the pattern of light and dark had been expected. After all, waves can interfere with one another. If two waves are superimposed so that crests match up with crests and troughs with troughs, the result is a single wave that is twice as high. If, on the other hand, crests are matched up with troughs, the two waves will cancel each other out. Since waves that bounce off a crystal in certain directions tend to reinforce each other, while waves bouncing in other directions cancel, a pattern of fringes results. Possibly, this sounds rather complicated. It isn't, really. Moreover, it is a very common phenomenon, one that can be observed with no scientific apparatus whatsoever. Anyone who wants to observe the interference of light waves can do so by looking at a light source between two fingers that are held slightly apart. If the fingers are then brought closer together, light and dark fringes will appear just before they touch.

In 1926, the Austrian physicist Erwin Schrödinger worked out the matter-wave theory in greater detail. He was able to discover how the waves propagated, and to apply them to the motion of electrons in atoms. For the first time, physicists had a theory which allowed them to work out the details of atomic structure.

It had been known since 1911 that an atom was made up of a

tiny, positively charged nucleus which was surrounded by negatively charged electrons. But no one had been able to give an adequate description of the manner in which the electrons orbited the nucleus. In 1913, the Danish physicist Niels Bohr had published a theory which had explained the behavior of the simplest atom, hydrogen. But Bohr's theory, which viewed electrons as particles, did not seem to work for helium atoms, the next most simple, or for anything that was more complicated.

Schrödinger's theory, called *wave mechanics*, did work. But it seemed to imply that the idea that an electron was a particle should be given up entirely. Electrons in atoms were described as packets of waves which spread out in all directions around the nucleus. If Schrödinger was correct, an electron had to be viewed as a cloud of negative charge that had no definite location.

At about the same time that Schrödinger propounded his theory, the German physicist Werner Heisenberg published his own theory of atomic structure. Heisenberg's theory, called *matrix mechanics* because it made use of abstract mathematical quantities called *matrices*, looked nothing like Schrödinger's. Strangely, it gave the same results.

It was soon demonstrated that even though Schrödinger's and Heisenberg's theories looked very different, they were mathematically identical. Today, physicists use the two versions interchangeably, and the term *quantum mechanics* is applied to both.

Even though the theories gave the same answers, they made use of two different pictures of the atom. Although Heisenberg had made no assumptions about what an atom looked like, his method seemed to fit in naturally with the particle concept. Schrödinger's theory, on the other hand, made use of the idea of waves. Neither physicist was easily reconciled to the other's approach. In a letter to the Austrian physicist Wolfgang Pauli, Heisenberg wrote, "The more I ponder about the physical part of Schrödinger's theory, the more disgusting it appears to me." Schrödinger was equally vehement. Writing of his adverse initial reaction to matrix mechanics, he said, "I was discouraged, if not repelled."

Meanwhile, physicists were discovering that neither version of quantum mechanics was so easy to interpret. True, they made it possible to perform calculations that could be verified by experiment. However, they failed to give a picture of the atom that was easy to understand.

Wave mechanics posed especially thorny problems of interpretation. The theory said that electrons were waves. But what, exactly, was the quantity that was oscillating? If the answer to that question was far from obvious, there was another problem that seemed even more baffling. It appeared that it was impossible to interpret Schrödinger's waves as waves in three-dimensional space. To be sure, the hydrogen atom was really not much of a problem; when one applied Schrödinger's theory to it, a three-dimensional wave equation resulted. However, a six-dimensional equation seemed to be required for the description of helium, which had two electrons. When there were even more electrons, things got even worse. When there were three electrons, the equations had nine dimensions. A ten-electron atom required thirty.

The extra dimensions did not prevent physicists from using the theory. A nine-dimensional mathematical quantity, for example, is not as fearsome as it sounds. The term only implies that there are nine mathematical coordinates, not that one must deal with nine dimensions of space. On the other hand, the presence of the extra coordinates made interpretation difficult. The presence of the extra dimensions seemed to imply that the waves could not be vibrations in ordinary space.

The problem of interpretation was soon solved by the German physicist Max Born. The waves were not real physical oscillations, Born suggested. On the contrary, they were *probability waves*. If one took the amplitude (i.e., height) of one of Schrödinger's waves and squared it, one obtained a quantity that gave the probability of finding the electron at any given point in space.

Perhaps an analogy will make Born's idea somewhat clearer. Suppose that one hides a small object in the sand near a beach. Suppose, further, that the wind blows the sand into dunes. Obviously, the higher a sand dune is, the greater the chance that it contains the object. Furthermore, the probability that the object will be found in any given dune depends upon the square of the height. If one dune is twice as tall as another, it will contain four times as much sand. The height of a dune has to be squared if one wants to get a good estimate of its volume, and hence of the probability that one will find the object in it.

Born's interpretation seemed straightforward enough. Moreover, it worked; it gave results that were consistent with experiment. But if anything, it made the wave/particle duality seem even more baffling. If one applied wave mechanics to a particle, the particle turned out to be a wave. But the wave did nothing more than describe the probability of locating a particle in a particular place.

Furthermore, there were experiments which indicated that the electron could sometimes manifest itself as a wave and sometimes as a particle. The electron diffraction experiments indicated that electrons were composed of waves that could interfere with one another. But it was possible to perform other experiments in which the particle character made an appearance. For example, when a moving electron strikes a fluorescent screen, a small flash of light will be produced. This is the principle upon which the picture tube of a television set is constructed. The only difference is that in a television picture tube, beams made up of very large numbers of electrons play back and forth across the screen.

There is nothing very wavelike about a particle that produces a flash of light in a particular spot. There are no interference fringes; the impact of a single electron is not spread out in space. The only way that one can interpret such an event is to view the electron as a particle. Even though one cannot see it, one can observe that some small object has collided with the screen.

Like light, electrons sometimes appeared to be waves, and they sometimes appeared to be particles. But whatever experiment was performed, they were always one or the other; they were never both at the same time. It was as though electrons (and light as well) decided to manifest themselves in one form or the other, depending upon what the experimenter was doing.

When an electron appeared as a particle, it seemed to lack a definite position or velocity. In this respect, it differed from the particles of classical physics. Quantum mechanics seemed to imply that if the position of an electron (or any other subatomic particle; quantum mechanics applies to them all) was precisely known, its velocity was completely undefined. On the other hand, if the velocity was measured, the idea of a precise position had no meaning.

This particular implication of quantum mechanics was discovered by Heisenberg; it is called the Heisenberg *uncertainty principle*. The uncertainty principle says that if we multiply the uncertainty in the position of an electron by the uncertainty in its momentum, the product is always larger than a certain small number. This means that the more accurately one quantity is known, the less we can tell about the other. When the uncertainty principle is stated in a precise manner, the two quantities position and momentum are used. However, momentum is just mass times velocity. So if the momentum is uncertain, the velocity is too.

Heisenberg gave an example which should make the meaning of the uncertainty principle a little clearer. Suppose we wanted to determine the momentum and position of an electron, Heisenberg asked. How would we go about it?

We could not get a very good idea of the position of an electron by looking at it in ordinary light, he pointed out. An electron, after all, is a very small object. Since the wavelengths of visible light are much larger than the electron itself, a very fuzzy picture would result, if one could obtain any picture at all.

To accurately determine the position of an electron, Heisenberg said, it would be necessary to look at it with something that had very short wavelengths: gamma rays, for example. Only then could one obtain a sharp picture. Unfortunately, illuminating an electron with gamma rays would destroy any information we had previously obtained about its momentum. A gamma-ray photon has so much energy that it would cause the electron to fly off at an unknown velocity in an unknown direction. If one tried to avoid this, the problem one had previously been trying to avoid would reappear. If one used red light, for example, knowledge about the electron's momentum would not be destroyed. But this would bring us back to the problem we had been trying to avoid in the first place; the wavelengths of red light are many times larger than an electron.

Nor does it help to use radiation of an intermediate wavelength: ultraviolet light, for example. Intermediate wavelengths would only partially destroy knowledge of momentum, but they would still give a moderately fuzzy picture of the electron's position. In this case, both momentum and position would be somewhat uncertain.

One should not misconstrue the uncertainty principle by assuming that an electron has a precise position and a precise momentum, which are altered when the electron is observed.* The only consistent way to interpret the principle is to assume that the electron has neither until an experiment that measures one or the other is performed. The particle/wave and position/momentum dualities are analogous. An electron, it seems, can appear to be something that has a position at some given instant, but no definite velocity. It can have a velocity, but no precise position. And it can exist in a state where both are somewhat indefinite.

At least this is the interpretation of quantum mechanics that is most widely accepted today. It is called the *Copenhagen interpretation,* because it was worked out in discussions between Bohr and other physicists at the Universitets Institut for Teoretisk Fysik in Copenhagen, an institution that was supported, in large part, by contributions from the brewers of Carlsberg Beer.

When physicists referred to the Institute, the long Danish name wasn't often used. Instead, it was known simply as "Bohr's Institute." The name was really quite appropriate. Bohr, the director, acted as a guiding spirit to an entire generation of physi-

---

* Strictly speaking, Heisenberg's example seems to imply precisely this. Heisenberg was guilty of engaging in oversimplification for the sake of clarity. Nevertheless, his picture, for all its defects, does allow one to gain some insight into the electron's behavior.

cists. During the 1920s and 1930s it was he who took the lead in attempting to understand what quantum mechanics meant. To this end, he was constantly inviting leading physicists to work at the Institute, or to spend some time there for the purpose of engaging in discussions.

These discussions led to the so-called *complementarity principle*, which was first enunciated by Bohr in 1927. I say "so-called" because complementarity seems to be more of an outlook than a principle. Bohr spoke of complementarity in a number of different ways, and he never gave an explicit definition of the term. This troubled some physicists. For example, Einstein complained in 1949 that he had never been able to find a clear-cut formulation of the concept, even though he had expended a great deal of effort on the task.

However, the idea of complementarity may not be so difficult to understand if one looks at it as a point of view rather than as a precisely defined physical law. The basic idea is a relatively simple one. According to Bohr, quantum mechanics could be understood only through the use of concepts which seemed to be mutually irreconcilable but which in fact were complementary to each other. An electron, Bohr said, was a wave, and it was a particle. It was not some hybrid of the two; it was both. Although "wave" and "particle" were mutually contradictory in the world of everyday experience, in the subatomic world they were not.

Bohr warned that any attempt to discard the wave and particle pictures as inapplicable to the subatomic world was misguided. One could not replace them with something new (which might, perhaps, be called a "wavicle"). Both had to be retained, because the only contact the physicist had with the quantum world was through his instruments of observation. The interactions of subatomic particles with these instruments could be described only in classical terms. Thus whenever the electron was "seen," it seemed to be either a wave or a particle.

If complementarity was nothing more than the idea that both poles of the wave/particle duality had to be retained, it would not be so difficult to define. However, Bohr applied the idea in a number of different contexts. For example, in the lecture in which he first stated the "principle," he stated that it was the concepts of "space-time coordination" and the "claim of causality" that were complementary.

Bohr pointed out that the state of an electron could be defined and treated causally only if it was left undisturbed. Observations inevitably changed its momentum or its position, and they changed other properties of an electron as well. If an electron was not observed, then one could not assign it a definite position in time and in space. But if one did observe an electron, one could not speak of causality. In other words, one could not predict where an electron would appear. Since its behavior was described by probability waves, one could speak only of the probability that it would manifest itself in a particular way.

In Newtonian physics, the behavior of an object was supposed to be perfectly predictable. If one knew the position and velocity of a planet or a cannonball or a speck of dust, and if one knew the forces acting on it, it was theoretically possible to predict its future behavior. In quantum mechanics, the situation was quite different. It was not a deterministic theory.

Quantum mechanics is one of the most successful theories in the history of physics. However, its predictions are statistical in nature. If one has a large number of atoms, or a large number of electrons, it is possible to state quite precisely what their average behavior will be. But one cannot tell what an individual atom or an individual electron is going to do. One can speak only of the *probability* that an atom will emit a photon of light at any given instant, one can speak only of the *probability* that an electron will appear at a certain place on a fluorescent screen, and one can speak only of the *probability* that a radioactive atom will decay within a certain period of time.

The quantum world, as Bohr pointed out, is a world of opposites. There is a sense in which the "claims of causality" do apply to subatomic particles like electrons. Probability waves propagate in a definite way. It is only when one disturbs the wave patterns by making an observation that the causality disappears. It is as though electrons materialized in a definite time and place only when one looked at them.

The American physicist Murray Gell-Mann has called quantum mechanics "that mysterious, confusing discipline, which none of us really understands but which we know how to use." When Gell-Mann says that physicists do not "understand" quantum mechanics, however, I don't think that we really need take him literally. He meant it in a deep philosophical sense. Quantum mechanics can be understood. However, it cannot be comprehended by the application of ordinary, commonsense concepts. As Bohr pointed out, subatomic particles exhibit properties that would seem to be contradictory if they were observed in a macroscopic object.

Quantum mechanics is more difficult to interpret than relativity. Einstein's two theories demonstrated only that certain commonsense ideas had to be modified. Quantum mechanics shows that, at least when we deal with happenings on the microscopic level, some of these have to be discarded entirely.

In fact, it is not even clear that subatomic particles have the kind of "objective reality" that we attribute to such objects as footballs and automobiles and planets. At least Bohr did not think that they possessed this quality. On one occasion, when his assistant Aage Petersen asked him whether he thought that the quantum-mechanical description mirrored some underlying reality, Bohr replied:

> There is no quantum world. There is only an abstract quantum physical description. It is wrong to think that the task of physics is to find out how nature is. Physics concerns what we can say about nature.

In other words, such objects as electrons did not "really exist." On the contrary, they were abstractions which helped physicists to codify their knowledge.

An electron, in Bohr's view, was not an object that had any intrinsic properties independent of experiment. On the contrary, it was a concept that helped one to explain why laboratory apparatus behaved the way it did. The electron did not have objective existence the way a macroscopic object did. Nevertheless, the concept of the electron as a real entity was indispensable if one was to have any hope of interpreting experimental data.

Bohr's view was a philosophical one, and not all physicists have accepted it. Nevertheless, it is a perfectly reasonable interpretation, one that is hard to avoid if one accepts the idea that the principle of complementarity is valid.

Chapter 2

SCIENCE, OR SCIENCE FICTION?

IN THE TELEVISION SERIES *Star Trek*, the spaceship *Enterprise* was able to travel between widely separated star systems in relatively short periods of time. The commander of the vessel could direct that it proceed at any of a number of different faster-than-light velocities. These were referred to as "warp 1," "warp 2," "warp 3" and so on.

It is not hard to guess what "warp" means in this context. It is nothing other than a shorthand version of the term "space warp," which has been used by science-fiction writers since the 1930s. The idea seems to be that if one can only "warp" space to a sufficient degree, then it is possible to circumvent the special theory of relativity and travel at velocities in excess of the speed of light.

The idea, which is loosely based on the concept of curved space in general relativity, is nonsense. There is nothing in general relativity which would imply that such a thing is possible. In fact, it is not even clear what the "warping" has to do with the speed at which a spaceship travels.

However, one cannot object to the use of such ideas in science fiction. After all, science fiction is the literature of the fantastic. If its authors commonly go beyond the bounds of reasonable scientific extrapolation, this only makes their work more interesting. It is no more reasonable to criticize the genre for mixing fact and fantasy than it would be to object to "Cinderella" on the grounds that a pumpkin cannot magically be transformed into a coach.

On the other hand, one can and should object when ideas that are nothing more than science fiction are presented as serious scientific possibilities. At times, science has found itself in the position of having to give serious consideration to ideas that seem, at first, to be contrary to common sense. But this does not imply that *any* crazy idea must be taken seriously. Sometimes attempts to understand natural phenomena force the scientific imagination to follow unexpected paths. But one cannot ignore observational data and reasonably well-confirmed theories. Wishing cannot make something true.

Science fiction sometimes masquerades as science. For example, it has been suggested that black holes might be used as time machines, or for interstellar travel. There has even been speculation about the possibility that black holes might serve as gateways to other universes. Such ideas are intriguing. Unfortunately, they all depend upon idealized mathematical descriptions of black holes which most likely have very little to do with reality.

Such speculations are based upon fanciful interpretations of *Penrose diagrams*, mathematical diagrams that were devised by the British mathematician Roger Penrose to elucidate the space-time structure of black holes. It so happens that if one draws a Penrose diagram for a black hole that is rotating, the black hole seems to be connected to a region of space that may not be part of our universe.

The Penrose diagrams seem to indicate that if an object which fell into a black hole followed a path that caused it to miss the singularity, it could enter this strange spatial region. Since the Penrose diagrams say nothing about what this region might be like, some scientists have speculated that it might be part of a different universe. Others have pointed out that alternative interpretations were possible. The object that fell into the black hole might reappear in our own universe at a different point in space or time. It was even possible, they said, that the object might fall back into our own universe at a point billions of years in the past.

If Penrose diagrams accurately describe black holes, then any of these things might be possible. In this case, travel to other universes, instantaneous travel in space or travel backward in time might be feasible.

If such ideas were accurate, they would have bizarre consequences for our understanding of the universe. If time travel were possible, strange paradoxes would result. For example, it would be possible for someone to travel back in time and to kill himself before he ever had the chance to make the trip. The problems associated with instantaneous travel across space are almost as bad. In fact, faster-than-light travel is time travel of a sort. It is one of the implications of special relativity that, in the reference frames of some observers, a faster-than-light vehicle would seem to arrive at its destination before it left its starting point.

The only possibility that does not create real problems is the suggestion that black holes might be gateways to other universes. If one defines "universe" to mean a region of space-time that is disjunct from our own, the idea is almost reasonable. In fact, speculation about other universes has become a respectable part of cosmological thinking in the last few years. For example, according to one recent theory, numerous universes came into existence at the same time, like bubbles in a bottle of tonic water.

However, there is no evidence that these other universes really exist. Their present status is roughly the same as that of the tachyon. There are no known laws of physics which would rule out the possibility of their existence. But of course, this is not a sufficient reason for believing in them. Since it is not likely that this evidence will be found in the foreseeable future, it is best not to take the idea that they might somehow be connected to black holes too much to heart.

The "other regions of space" that the Penrose diagrams describe most likely correspond to nothing real. It is probable that they are

48 DISMANTLING THE UNIVERSE

nothing more than a mathematical artifact resulting from a description that is too simplified.

One of the problems associated with any theoretical investigation of the structure of black holes is the fact that all such investigations must make use of Einstein's general theory of relativity. However, it is known that general relativity has limits. Specifically, it is no longer valid under the conditions that prevail near the singularity of a black hole. In order to describe matter in the singularity, one would need a theory of *quantum gravity*, a theory that combined quantum mechanics and general relativity. Matter in the singularity would be so compressed that quantum effects would become important. Its behavior could not be described by relativity, which deals only with gravitational forces. Unfortunately, no one seems to have any idea how a quantum-gravity theory might be constructed.

It could very well be that this does not invalidate the Penrose diagrams. After all, the diagrams say that an object gets into another universe by following a path that causes it to miss the singularity. It may be that general relativity would give an accurate description of the behavior of such an object.

However, there is another sense in which the Penrose diagrams are an idealization. When one draws such a diagram, the assumption is made that all the mass in a black hole is concentrated in a singularity. But this may be true only in the first approximation. If we assume that general relativity can be applied to the interior of a black hole, we have to conclude that the gravitational fields near the singularity would be so intense that large quantities of matter and antimatter* would be created from empty space. Some of the gravitational energy in the black hole would materialize as mass. It is extremely difficult to tell just what the effects of this would be; the equations of general relativity are not easy to solve, even in relatively simple situations. However, there are reasons for suspecting that the paths to the "other regions of space" which the diagrams describe would disappear.

* The properties of antimatter will be described in detail in Chapter 8.

At present, physicists are not absolutely sure that black holes exist. The circumstantial evidence is very strong. At some time in the not-too-distant future, we may very well obtain evidence that is conclusive. However, if our technology ever advances to the point where travel to black holes is possible, I seriously doubt that we will see spaceships flying into them in an effort to travel to other universes. Any ship that did so would almost certainly be destroyed shortly after it penetrated the event horizon, if it managed to survive that long.

Almost anything is possible in our imaginations. But it does not follow that almost anything is possible in reality. When we are forced to adopt strange new ideas, that is science. When we try to see what strange ideas we can possibly entertain, that is science fiction.

Science fiction can be entertaining. Consequently, one encounters it everywhere. It crops up in supposedly "factual" magazine articles, in newspapers, on television and sometimes even in scientific discussions. Science fiction has become so ubiquitous that we even encounter situations in which science fiction is used to explain science fiction.

For example, it has been suggested that the gravitational distortions of the geometry of space and time that are predicted by general relativity might provide an explanation for extrasensory perception. It is postulated that there is something called a "biogravitational field" which can distort "local subjective space-time," and thus make such phenomena as clairvoyance and telepathy possible. Alternatively, it has been suggested that tachyons might provide a possible mechanism for telepathy, and that ESP might somehow depend upon events which take place at the "subquantum level."

The only thing wrong with these explanations is that things which probably do not exist are used to explain a phenomenon which most likely does not exist either. There is no evidence which would indicate that there is any such thing as a "biogravitational field"—or tachyons, for that matter. Nor is it easy to understand what the words "subquantum level" are supposed to mean. Finally, there is really not any good evidence that extrasensory perception is real.

Parapsychologists often claim that the existence of ESP has been conclusively established. They claim that one no longer need perform experiments to demonstrate its reality. The only problem that remains, they say, is that of understanding its properties.

The skeptics, however, are not convinced. In their view, the evidence for the existence of ESP is anything but conclusive. They maintain, furthermore, that research in parapsychology is frequently associated with shoddy methodology, and is characterized by overly credulous attitudes on the part of the investigators.

The debate is likely to continue for quite some time. However, the skeptics seem to have the better arguments at the moment. The more closely one looks at parapsychological research, the less convincing its results seem to be. Although ESP experiments have been performed for more than half a century, the existence of the phenomenon being studied does not seem to have been established.

Scientific experiments are supposed to be repeatable. It should not make any difference who it is that conducts an experiment as long as he is a reasonably competent investigator. This does not seem to be true in the case of ESP. For more than thirty years, skeptical psychologists have been attempting to duplicate the results reported by the parapsychologists without much success. They have reported one negative result after another.

It may be that doubt somehow acts to inhibit extrasensory powers, and there are at least one or two cases where skeptics have become believers, so perhaps this lack of repeatability is not entirely conclusive. Nevertheless, it does seem to indicate that the case for the existence of extrasensory perception is not so strong as has been claimed.

When one looks at the experiments performed by the parapsychologists themselves, certain peculiar facts begin to emerge. It seems that many of the early ESP experiments were performed under rather lax conditions. Few precautions were taken to prevent deception by the subjects or unconscious errors by the experimenters. When the experimenters tightened up their procedures in response to criticism, ESP scores exhibited a mysterious tendency to decline. It became apparent that the more conscientious one was about setting up a rigorously controlled experimental setting, the less likely it was that any significant results would be obtained.

To make matters worse, the field of parapsychology has been plagued by revelations of fraud. Subjects who had been thought to possess extraordinary extrasensory powers sometimes turned out to have been cheating. Furthermore, it has been discovered that some prominent parapsychologists have resorted to cheating themselves. One of these cases, which came to light in 1974, involved fraudulent experiments conducted by Walter J. Levy, Jr. Three fellow staff members of the Institute for Parapsychology at Duke University (of which Levy was then director) caught him falsifying test results in an experiment on rats. Levy had been attempting to demonstrate that rats could use ESP to influence electrical apparatus. At the time, the Institute had long been considered the most important center for parapsychological research.

In recent years, parapsychologists have taken to studying the powers of individuals who claim to be psychics. As a result, parapsychology as it is practiced today often bears a strong resemblance to the "psychical research" that took place around the turn of the century, when investigators worked with mediums who claimed to be able to establish contact with the spirit world. Not surprisingly, some of the psychics studied by parapsychologists have been exposed as fakes, or have had their tricks duplicated by professional magicians. For example, according to author Martin Gardner, the American stage magician James Randi now does an even better job of bending keys than the Israeli psychic Uri Geller.

None of this proves, of course, that all psychics are fraudulent. If one thoroughly investigated ninety-nine psychics and successfully demonstrated that all of them were fraudulent, this would not prove that the one hundredth was not genuine. For that matter, it wouldn't prove that the first ninety-nine cheated all of the time. After all, they could still claim that they resorted to fraud only on those occasions when their powers failed them.

Such an argument was actually used by believers in the powers of the Italian medium Eusapia Palladino in the late 1890s. It seems that Palladino was caught cheating when she conducted seances in Cambridge, England, in 1895. But this seems not to have fazed her supporters in the least. They claimed that she was only 60 percent fraudulent. She used tricks, they said, only when the real spirits would not make an appearance.

It is practically impossible to demonstrate that a phenomenon does not exist. As a result, anyone who wants badly enough to believe in ESP generally has no trouble finding a way to do so. However, if extrasensory perception is real, it must be a very strange phenomenon. After all, it would be hard to think of anything else that can be detected only on those occasions when skeptical investigators are not looking.

It seems reasonable to conclude that both ESP and the phenomena that have been invoked to explain it are science fiction. The reasons for believing that extrasensory perception exists appear to be no better than those for believing that the spaceship *Enterprise* can travel at "warp speed" throughout the universe.

Science fiction sometimes becomes scientific fact. The French poet Cyrano de Bergerac wrote a story about a voyage to the moon in the seventeenth century. A science-fiction story about an atomic bomb was written years before the first nuclear device was exploded in Alamogordo, New Mexico. Can one be sure, then, that the reality of ESP will never be demonstrated?

Perhaps not. However, it does seem rather unlikely. It seems equally unlikely that we will ever discover that black holes are gateways to other universes, or that it is possible to travel into the past. It is possible to believe in practically anything. However, the vast majority of the concepts that the human mind is capable of forming have little to do with objective reality. The discovery that it is possible to imagine something does not imply that it is real. The rational skeptic need not demand overwhelming proof before he is willing to entertain an idea. However, he is likely to ask for some modicum of evidence.

Sometimes the dividing line between science and science fiction is a fine one. In fact, it is sometimes necessary to investigate a whole series of outlandish ideas before one hits upon the solution to a scientific problem. The very same scientists who remain skeptical about such phenomena as ESP habitually consider notions that are even more fantastic.

There is nothing paradoxical about this. It is one thing to believe in the powers of psychics because that belief is emotionally gratifying. It is quite another to entertain audacious ideas in the hope that they might give the answer to a scientific question.

When Louis de Broglie postulated the existence of matter waves, it seemed to many scientists that he was engaging in a fantastic kind of speculation. If the term "science fiction" had been in common use at the time, it might very well have been applied to De Broglie's thesis.

Nevertheless, there is a difference between De Broglie's speculative idea and some of the ideas that we have been considering. De Broglie was making an attempt to discover a road that would lead to the solution of certain puzzling scientific problems. When he began to work on his theory, scientists had not found a theory that would adequately explain atomic structure. De Broglie had ample justification for advancing a speculative idea. Even if his theory had turned out to be wrong, it would have had some value. In that case, physicists would have been aware that there was an approach that didn't work.

On the other hand, speculation about bizarre properties of black holes and about the use of ESP by psychics does not have this kind of motivation. If we were to encounter extraterrestrial beings who claimed to have journeyed from another universe through a black hole, there would be good reason for looking into the possible implications of Penrose diagrams. If the psychics' predictions that are published in *National Enquirer* were consistently accurate, there might be some reason for asking whether "biogravitational fields" could possibly explain ESP. But as long as they are not accurate, such ideas can safely be ignored.

There is another kind of borderline science fiction. One can hardly even call it "science fiction" without hastening to add that it is much more respectable than speculation about possible mechanisms for ESP. What I am referring to is speculation that mixes science and philosophy. Although numerous eminent philosophers and scientists have engaged in speculation of this sort, they cannot be said to be doing "scientific" speculation when they go beyond the bounds of present knowledge.

It is often stated, for example, that the indeterminacy of quantum mechanics leaves room for human free will. The idea seems to be that if subatomic events are subject to chance, then the functioning of the human brain must be indeterministic in some sense also. It is this lack of determinism which supposedly makes free will possible.

It is obvious that the question whether free will exists is still of a philosophical nature. Science does not yet know enough about the structure or the functioning of the human brain to say anything about the problem that is very meaningful. Although knowledge about neural processes is rapidly increasing, no one really has any idea as to how these processes might relate to such things as "free will" or "mind" or even "consciousness." It is not even obvious why consciousness should exist.* After all, it is not clear why evolution did not design us as complicated automatons rather than as beings with self-awareness.

If biology cannot tell us very much about free will, then it would be very surprising if physics could. It is possible that physics will eventually be able to explain all of chemistry, and that chemistry will explain biological phenomena. However, the reduction of biology to physics has not been carried out in practice, and there is very little chance that it will be in the foreseeable future.

---

* It has been suggested that consciousness is analogous to the software of a computer. However, this analogy does not really explain very much.

Quantum mechanics does not tell us anything very meaningful about the functioning of a biological cell. Hence the idea that it can explain free will, which is something that presumably operates at a higher level, must be considered rather improbable. For that matter, one can't be sure that free will exists. We all have subjective feelings which tell us that our choices are free. However, not all philosophers agree that these subjective feelings are very meaningful. It is perfectly possible to advance the opinion that all human behavior is rigidly determined.

The history of the notion that there is a connection between quantum indeterminacy and free will is a curious one. It is sometimes said that the idea was originally promulgated by the British astronomer Sir Arthur Eddington in the early 1930s. But Eddington made no such suggestion. In fact, he stated quite explicitly that the idea was "nonsense." In Eddington's opinion, the assumption of free will implied that the brain must contain "conscious matter" which correlated processes taking place in the relatively inert matter that made up the remainder of the nervous system. If free will did operate, Eddington said, it did so without regard to quantum indeterminacies. At best, chance events on the subatomic level could introduce chance elements into behavior. But chance elements did not make human will free; they had nothing to do with the idea of free choice.

It should be obvious that speculation of this sort is philosophical, not scientific. "Conscious matter" is nothing more than a new name for "mind" or "soul." But Eddington realized that his speculations were philosophical in nature. He spoke of consciousness as something that was "outside physics." When he introduced the notion of "conscious matter," he realized quite well that he was discussing something that had never been seen in a laboratory.

Today, many authors allow themselves to be somewhat less rigorous when they discuss the problem. Sometimes they state that quantum indeterminacies make free will possible without even giving any arguments in support of the hypothesis. Sometimes they engage in outright fantasy and speak of connections between quantum events and consciousness. In his book *Taking the Quantum Leap*, physicist Fred Alan Wolf goes so far as to speak of "atomic minds" that correlate and come together into "one mind knowing."

It is not very surprising that scientific and philosophical questions should sometimes become entangled. After all, science was once a branch of philosophy. In Newton's day, for example, scientists were called "natural philosophers." After the two fields separated, the concerns of one sometimes continued to be the concerns of the other. Today, one of the most active fields of philosophical inquiry consists of attempts to clarify the logical structure of science.

In particular, the philosophical question of the existence of free will has been entangled with scientific speculation for a long time. One of the reasons that so many authors speak of quantum mechanics and free will today is that two hundred years ago, some scientists were wondering if the laws of physics didn't imply that free will was *not* possible.

At one time it was thought that physics implied a universal determinism. The best-known version of the argument is that of the eighteenth-century French mathematician and astronomer, the Marquis Pierre Simon de Laplace, who stated that:

> An intelligence knowing, at any given instant of time, all
> forces acting in nature, as well as the momentary positions
> of all things of which the universe consists, would be able to
> comprehend the motions of the largest bodies of the world
> and those of the smallest atoms in one single formula,
> provided it were sufficiently powerful to subject all data to
> analysis; to it, nothing would be uncertain, both future and
> past would be present before its eyes.

Laplace did not believe that any such intelligence (or "Laplace demon," as it later came to be called) existed. But that wasn't his point. What he was saying was that if the laws of Newtonian physics were valid and exact, then every future configuration of everything in the universe was determined. Obviously, the calculation that Laplace envisioned could not be performed. However, that did not alter the fact that everything, from the smallest atom to the largest astronomical body, acted in accord with the laws of physics. The behavior of any particle depended only upon the forces that acted upon it. If one could only determine what the forces were, then one could predict the future behavior of the particle for all eternity.

If human bodies are made up of the same materials that constitute inanimate matter, then it seems to follow that this Laplacian determinism would apply to human actions also. If the atoms that make up a body behave according to deterministic laws, then one must conclude that the body as a whole is subject to determinism also.

It is now realized that the laws of physics are approximations. No law describes the working of the universe with perfect accuracy. At best, any law explains natural phenomena within the limits of some small margin of error.

In Newton's day, it was believed that the inverse-square law of gravity was exact. Every body in the universe attracted every other body with a force that was inversely proportional to the square of the distance. But when Einstein propounded his general theory of relativity, it was realized that Newton's law was nothing more than an approximation. As long as the gravitational forces under consideration were not too intense, Newton's law worked extremely well. But the reason that it worked well was not that it was exact. The reason was that the corrections which Einstein's theory made necessary were so small that it was practically impossible to measure them.

When gravitational fields become strong, the inaccuracies inherent in Newton's law increase, and it becomes necessary to use Einstein's theory instead. But the general theory of relativity is only an approximation also. When gravitational fields become extremely strong, for example near the singularity of a black hole, it breaks down. Nor is it likely that a theory of quantum gravity would provide us with exact predictions either. If such a theory were found, it would most likely become obvious that it too had its limits.

One of the reasons that physics has made so many advances during the twentieth century is that physicists have been observing nature under ever-more-extreme conditions. They have peered into the realm of the very small in order to study the behavior of subatomic particles. Astronomers have found ways to look billions of light-years into space. As they have done so, they have been able to see billions of years into the past. Since a light-year is the distance that a ray of light will travel in one year, an object 5 billion light-years away is seen by light that it emitted 5 billion years ago. Other branches of physics involve the study of nature under conditions that could not even be created a few decades ago. For example, matter can now be cooled to temperatures close to absolute zero.*

Very often it is discovered that the existing laws of physics are not adequate to describe nature under such conditions. When this happens, physicists must find ways to discover laws that are better approximations than the old ones. If their approximations are good enough, they will be able to make predictions that can be confirmed within the limits of experimental accuracy. But if the new laws are more accurate, it certainly does not follow that they are exact. The concept of an exact law of nature is an abstraction. There are only laws that are reasonably accurate under most conditions, and those that are a little better.

It is sometimes said that quantum mechanics pokes holes in Laplace's argument for determinism. Perhaps it does. But even if quantum mechanics had never been discovered, the argument would not be very convincing. It lost all its cogency once physicists discovered that they were not uncovering precise laws of nature. Determinism is not a consequence of Laplace's argument; it is an implicit assumption.

The only reasonable conclusion that one can make is that physics has nothing to say about the determinism/free-will problem. It was a philosophical question two thousand years ago, and it is still a philosophical question today. Quantum mechanics no

* Absolute zero is the lowest possible temperature. It is the temperature at which there is no heat, and no molecular motion.

more makes free will possible than "biogravitational fields" allow psychics to predict earthquakes.

Should we conclude, then, that quantum indeterminacies have no effect on processes that take place in the human brain? Not necessarily. This is not the same question. It could be that chance quantum events introduce chance elements into brain functioning and into human behavior. It is not very likely. The brain appears to be "wired" together in such a way as to ensure that chance happenings will not affect it much. Connections between neurons are set up in such a way that there is a great deal of redundancy. A nerve impulse can generally get from one section of the brain to another along many different paths. If some chance event interrupts the impulse along one of these pathways, it will continue to be propagated along many others.

Every day, approximately 10,000 cells in the brain of an individual die, and are not replaced. The redundancy ensures that these cell deaths will not have any significant effect. Under such conditions, it is hard to see how chance events that cause a cell to fire, or not to fire, can significantly affect behavior. If a "large" event, such as the death of a large number of cells, doesn't change things significantly, it is hard to imagine that a "small" one would.

Quantum events could conceivably affect the firing of some neurons. A typical brain cell is connected to numerous other cells by *synapses*, or junctions. A neuron may receive impulses from something like a thousand other cells; it may transmit impulses to a thousand or even ten thousand others. It is perfectly conceivable that random quantum events could facilitate or prevent the transmission of an impulse at any given synapse. But in view of the brain's redundancy, it is not obvious that there would be any effects other than the creation of a small amount of "noise" or "static." If random events caused a cell to fire, or not to fire, most likely the result would be nil. A brain is not constructed like a pocket calculator or a television set, which may act oddly or fail to operate if one component begins to behave erratically.

Quantum mechanics does not have anything significant to say

about the determinism/free-will problem because physics never had much to say about free will or determinism in the first place. The Laplacian argument that the universe is deterministic doesn't work because it depends on the assumption that exact laws of nature exist. But physics knows of no exact laws; it deals only with approximations. Quantum mechanics does not overturn determinism because there was nothing to overturn.

When we pass from general arguments about determinism to specific arguments concerning the structure of the brain, it is difficult to turn up anything that seems very significant. At best, quantum events would cause the firing of a few random neurons. It is hard to see that this could have anything to do with free will. In fact, free will seems not to be related to randomness at all. As the German philosopher Ernst Cassirer pointed out, the concept of free will implies ordered control of one's actions.

At present, physics has nothing to say about what we call "free will," "mind" and "consciousness." Biology has very little to add. But it does not follow that science will not be able to provide answers to some of these questions in the future. After all, philosophical questions become scientific ones whenever science advances to the point that it can give us the knowledge to deal with them. In Aristotle's day, questions concerning the nature of matter and the nature of the cosmos were considered to be part of philosophy. Today, they are thought to be scientific in nature.

If science does eventually tell us anything about the nature of free will, it is likely that we will be in for some surprises. As the stories of the discovery of relativity and quantum mechanics demonstrate, the results of science are often more astonishing than the speculations associated with science fiction. When a scientific revolution takes place, scientists discover that they must dismantle the conceptual universe that they have inherited, and reassemble it anew. Science fiction generally does nothing but add arbitrary new features to an existing picture of the universe.

The rather nonsensical conception of "space warps" that science-fiction writers used to use represented nothing more than an attempt to make the universe seem more interesting by engaging in speculation that had little basis in fact. The bizarre implications of Penrose diagrams are of a similar nature. So are speculations about extrasensory perception. If one wants to believe in ESP, it is not necessary to make any great changes in one's conceptual outlook. One need only assume that human beings can obtain information in other ways than through the usual channels of sensory awareness. Finally, speculation about quantum mechanics and free will requires no revolutionary changes in one's outlook. It consists of nothing more than attempts to find connections between a well-confirmed scientific theory and an old philosophical problem.

On the other hand, science has changed our conception of the universe in dramatic ways. Einstein showed, in his general theory of relativity, that it was necessary to give up centuries-old ideas about the nature of space and time. Quantum mechanics demonstrated that the concepts which described the world of everyday experience were no longer applicable on the subatomic level. An electron could be a wave and a particle. It could "exist" in an abstract sense, and yet lack such properties as a precisely defined momentum and a precisely defined position. It turned out that even when one had complete knowledge about an individual "particle," its behavior was unpredictable. When quantum mechanics was discovered, physicists were forced to conclude that it was not always very meaningful to speak of the properties of an individual electron or an individual atom. One had to make do with statistical predictions about the behavior of ensembles of particles instead.

STYLE AND
SCIENTIFIC DISCOVERY

WHEN EINSTEIN WAS A SIXTEEN-YEAR-OLD BOY, he began to wonder what would happen if one could travel at the speed of light. He quickly observed that there was something very puzzling about the idea, for if one could travel at such a velocity, light itself would disappear.

The year was 1895. Physicists had known for some time that light was a form of electromagnetic radiation; it consisted of oscillating electric and magnetic fields. When these oscillations struck the retina of the eye, or interacted with a piece of laboratory apparatus, light was observed.

Einstein realized that if one could move at the velocity of light, the oscillations would no longer seem to exist. They would become frozen. Where a stationary observer would see a series of alternating crests and troughs as a light wave moved past him, an observer who was traveling at the same speed as the light wave would see only a motionless crest, or a motionless trough, and nothing else. Einstein saw that there was something very paradoxical about such an idea. Motionless "oscillations" simply did not exist in nature. Consequently, he began to wonder if velocities equal to that of light were possible.

This adolescent thought experiment suggested to Einstein that

motion at velocities equal to that of light was not possible. It may
have motivated him to seek a theory which explained why such a
thing could not happen. However, it does not explain how Einstein happened to think of the specific postulates upon which
the special theory was based. It no more explains the genesis of
his theory than the anecdote about Newton and the apple* explains the discovery of the inverse-square law of gravitational
attraction.

How do scientific discoveries come about? Does the theoretical
scientist depend primarily on logical deduction? If Einstein had
never been born, could the special theory of relativity have been
discovered by any physicist who was acute enough to realize that
the idea of travel at the speed of light was paradoxical? Or does
theoretical discovery depend upon a kind of creative imagination
similar to that of the artist or the poet?

The question of how Einstein happened to think of the special
theory of relativity has no obvious answer. It is a topic about
which scholars still argue. However, the other questions are not
so difficult to answer. It appears to be fairly obvious that most
significant discoveries depend both upon logical thinking and
upon creative leaps of imagination.

There are a few relatively "pure" examples of one or the other.
Planck's discovery of the quantum theory of light seems to have
depended primarily upon deduction. Planck began by deriving a
mathematical formula that would accurately describe blackbody
radiation. Next, he attempted to see what the theoretical consequences of the formula were. Although intuition must have
played a role of some sort, it was probably a relatively minor
one.

On the other hand, De Broglie's hypothesis about matter waves
was so much of an intuitive leap that it was practically a shot in
the dark. At the time, there was no experimental evidence which
would have indicated that particles would exhibit wave charac-

---

* This particular story, by the way, seems to be more accurate than the
majority of such anecdotes. It appears that Newton did begin to think about
gravitation after watching an apple fall from a tree.

teristics. De Broglie's theory was not the result of deduction; it was an inspired guess.

The examples of Planck and of De Broglie are rather atypical, however. In most cases, both logical analysis and creative imagination play important roles in scientific discovery. New theories rarely come about because one scientist or another has expended a great deal of effort analyzing experimental data. They just as rarely depend upon insight alone. Typically, scientific discovery is a two-part process. The first thing that happens is that a scientist experiences a sudden insight. Then, if he is lucky, he finds that the insight has logical consequences that will clear up an outstanding scientific problem, or explain baffling experimental results.

I say "if he is lucky" because not every scientific idea is a good one. Insights can be accurate, and they can be totally misleading. One could even go so far as to say that the majority of the scientific theories that have been proposed, or thought of, have been wrong, just as the majority of the poems that have been written and the majority of the pictures that have been painted were not very good. Great scientists and great artists sometimes have bad ideas. Mediocre scientists and artists have them frequently.

There are more parallels between science and the arts than immediately strike the eye. Like artists, scientists often have unique styles. Furthermore, their ideas of what a good scientific theory should be like are strangely reminiscent of artistic convictions. Admittedly, the concept of "style" is not often applied to the sciences. Nevertheless, when one looks at the work of Planck, of Bohr and of Einstein, it is difficult to avoid being struck by the differences between them. It is apparent that these three men attempted to understand the universe in three stylistically very different ways.

Planck was a conservative classical physicist. Although he was the instigator of the twentieth-century revolution in physics, he was a reluctant revolutionary. He was the originator of the quantum theory, yet he was so committed to classical nineteenth-century ideas that he spent ten years of his life trying to find a way to do away with the concept of quanta entirely.

It wasn't a desire for innovation that led Planck to the discovery of the quantum emission of light. He made the discovery because he perceived that the blackbody problem represented a glaring gap in scientific knowledge. He labored for years to plug that gap, and finally proposed that light was emitted in packets only because he could see no way to avoid that conclusion.

It would be a mistake to view Planck as an old fuddy-duddy. He was a brilliant, imaginative physicist. He was the first to appreciate the significance of Einstein's special theory of relativity. He was on hand when Einstein's rather unconventional paper arrived at the offices of the journal *Annalen der Physik*, and he defended Einstein's theory in print the following year. Yet although he understood Einstein's imaginative ideas, he distrusted his own insight.

Relativity, after all, represented an extension of ideas that had been accepted for a long time. When Einstein propounded the special theory, he reached some surprising conclusions. But he did not break with the old ideas the way Planck did when he formed the quantum hypothesis. Interestingly, Planck, who endorsed relativity with enthusiasm, remained skeptical when Einstein proposed that light traveled through space as photons. Above all, Planck seems to have wanted to preserve the concepts of classical physics.

If Planck perceived the laws of nineteenth-century science as absolute truths that could be expanded upon but not overthrown, Einstein was willing to question anything in the structure of science. By proposing his theory of special relativity, he showed that he was willing to make audacious assumptions. In his paper on quantum theory, published the same year, he suggested that light was made up of particles, even though he knew that there was conclusive evidence that it was a wave phenomenon. When he endorsed De Broglie's matter-wave hypothesis, he showed that he was willing to consider an idea that most physicists would have rejected out of hand.

In Einstein's view, only one thing was important. Physics had to be made into something that was logical and coherent. If his quest for an inner logic led to ideas that seemed outlandish, Einstein would happily conclude that the world was stranger than it seemed to be. It seems never to have occurred to him that a quest for logical structure could lead to conclusions that were incorrect.

Planck and Einstein also differed in the emphasis they placed on experiment. Planck began to work on the blackbody problem because he saw that theory and experiment did not agree. On the other hand, Einstein paid little attention to experimental results. Like the ancient Greek philosophers, he believed that pure thought could grasp the structure of reality. If a theory was elegant and logically compelling, then it had to be true. If experiment indicated otherwise, then so much the worse for experiment.

In 1906 the German physicist Walter Kaufmann published the results of a long series of experiments in which he had measured the mass of moving electrons. His results agreed with some theories and disagreed with others. In particular, they failed to substantiate the predictions of the special theory of relativity. There was a small but significant difference between Kaufmann's findings and Einstein's calculations.

But Einstein was not troubled. Commenting on the two theories that Kaufmann's results did support, he wrote, "In my opinion both theories have a rather small probability because their fundamental assumptions concerning the mass of moving electrons are not explainable in terms of theoretical systems which embrace a greater complex of phenomena."

In other words, no matter what the experiments said, the competing theories could not be true because they did not fit into far-reaching, clear-cut theoretical patterns. In Einstein's eyes, the logical structure of a theory was more important than results that were obtained in the laboratory.

Einstein became even more dogmatic about the primacy of theory over experiment when he heard of the results of an experimental test of the general theory of relativity. In 1919 a group of astronomers headed by Arthur Eddington went to Africa to observe the path followed by starlight that grazed the surface of the sun during a solar eclipse. Einstein's theory had predicted that the starlight would be deflected by a certain amount. His prediction of the bending of starlight in the sun's gravitational field could be checked only during a total eclipse because at other times the light of the sun blotted out the images of nearby stars.

Although the observations were difficult, Eddington managed to obtain results that confirmed Einstein's theory within the limits of experimental error. One might expect that Einstein would have been overjoyed when he heard of this. But he wasn't; he was relatively unmoved. When one of his students, Ilse Rosenthal-Schneider, asked him why he did not seem to be as excited as she was, Einstein replied, "But I knew that the theory is correct." When she asked him how he would have reacted if the theory had not been confirmed, he said, "Then I would have been sorry for the dear lord—the theory is correct."

Einstein did sometimes make mistakes. However, his greatest error was not the result of paying too little attention to observational data. It came about when he paid too much attention to them.

In 1917, Einstein set himself the task of using his general theory of relativity to find equations that would describe the structure of the universe. To his surprise, he found that his theory indicated that the universe had to be in a state of either contraction or expansion.

In 1917, all the astronomical observations that had ever been made seemed to indicate that the universe was static. It had been known for thousands of years that the constellations did not change in size or appearance. The heavens seemed to be unalterable; one spoke of the "fixed stars" to distinguish them from the moving planets. It had apparently not yet occurred to anyone that the universe might evolve on a time scale that was much longer than the time spans by which events were measured on earth. No one realized that the universe might be changing, that the only reason it appeared static was that it was seen in slow motion.

Einstein didn't realize this either. Rather than follow where his theory led him, he accepted the conclusion that the universe must be static. Since his equations indicated that it could not be, he decided to "fix them up." Einstein accomplished this by introducing a quantity that he called the *cosmological constant*. This constant was the mathematical representation of a force that balanced out the effects of gravity at large distances. Once this force was "put in," a static universe became mathematically possible.

Twelve years later, in 1929, the American astronomer Edwin Hubble discovered that the universe was in a state of rapid expansion. Hubble's observations indicated that the light from distant stars was reddened. This *Doppler shift*, or *red shift*, indicated that distant galaxies were rushing away from the earth. Since it was ludicrous to assume that the earth occupied any special position in the center of the universe, only one conclusion was possible. Galaxies and clusters of galaxies were moving apart. The universe was expanding.

It wasn't long before Einstein realized that if he had only trusted his own theory, he could have predicted this result twelve years before Hubble discovered it. By 1931, Einstein had rejected the concept of a cosmological constant. In later years he was to speak of the introduction of the quantity as "the greatest blunder of my career."

But the story does not end here. After Hubble had made his discovery, some cosmologists began to wonder if there might not be good reasons for retaining Einstein's constant after all. It seemed that Hubble's results gave an age for the universe that was much too young. The cosmologists observed that if the universe was expanding, then there must have been a time when all the matter in it was compressed together. If one took Hubble's figures and worked backward, it was possible to calculate that approximately 2 billion years had passed since the expansion began.

However, this result did not agree with determinations of the age of certain terrestrial rocks. Radioactive dating methods showed that some of these rocks were 3.5 billion years old. This contradicted the 2-billion-year figure for the age of the universe.

It was absurd to assume that rocks found on the surface of the earth could be older than the universe itself.

There seemed to be only one way out of the dilemma. If one reintroduced Einstein's cosmological constant into the calculations, the theoretical age of the universe could be extended. If there existed another long-range force besides gravity, the expansion would proceed at a different rate. If the cosmological force was of the right strength, it was possible to obtain an age for the universe that was greater than 3.5 billion years.

But Einstein refused to accept this as a way out. He was now convinced that the introduction of the cosmological constant had been a mistake, that it created a flaw in the logical structure of his theory. It should not be brought back in merely to bring theory into agreement with astronomical data. Although Einstein could suggest no method of resolving the contradiction, he remained adamant. He had blundered once, and he was not going to again.

Einstein's obstinacy seems almost pigheaded. And yet he turned out to be right in the end. During the 1950s, two American astronomers, Allan Sandage and the German-born Walter Baade, demonstrated that there had been systematic errors in Hubble's distance measurements. When Hubble had done his work, astronomical knowledge—at least in some areas—had been in a relatively primitive state. Lack of knowledge had caused Hubble to confuse certain kinds of stars that were really quite different from each other. Since he had used the brightness of these stars to calculate the distances to nearby galaxies, his results for the expansion rate had been inaccurate. And if the expansion rate had been determined incorrectly, the estimate for the age of the universe would be wrong too.

Sandage and Baade corrected the age of the universe to 10 billion years. Although this was still a bit too small—the modern estimate is 18 billion—it was large enough to account for 3.5-billion-year-old rocks. It became apparent that Einstein's intuition had not failed him. There was no discrepancy, and the cosmological constant did not have to be brought back in.

Until Einstein reached the age of thirty-five or forty, he seemed to have an almost clairvoyant insight into the workings of nature. When Einstein decided that a theoretical idea "felt right," it almost invariably turned out to be true. He seemed to have the mysterious ability to penetrate beyond the world of appearances and to see, in his mind, the hidden laws that explained natural phenomena.

Einstein often spoke in a manner that would have seemed more appropriate for a mystical philosopher, or for an Eastern seeker after enlightenment, than for a Western scientist. When he talked of his attempts to understand the laws of nature, he often expressed himself in enigmatic ways. On one occasion, he spoke of his attempts to discover the "secrets of the Old One." "God is subtle, but not malicious," he said on another, when he wanted to express the idea that although natural laws were sometimes difficult to discover, they were never incomprehensible.

Perhaps an Eastern mystic would not have used the words "Old One." But when Einstein did so, he was not speaking of God as though He were an individual. Einstein had lost his belief in the personal God of Judaism and Christianity while he was still an adolescent. When he spoke of "God," he was simply referring to the logical patterns that were observed in nature. In Einstein's eyes, the natural universe partook of the divine. Perhaps this idea can best be expressed in Einstein's own words. In 1929, in response to an inquiry from a Rabbi Goldstein in New York, Einstein stated that he believed in "Spinoza's God who reveals himself in the harmony of all that exists, not in a God who concerns himself with the fate and actions of men."

One should not conclude that Einstein was infallible. He very definitely wasn't. In fact, after he passed the midpoint of his life, his intuition began to fail him. He spent his last forty years in an attempt to discover a *unified field theory* that would combine the laws of gravity and electromagnetism. Today, it is apparent that Einstein was following a blind alley. Although there is hope that a unified theory might be possible, it has become apparent that the theoretical methods employed by Einstein are not likely to produce one.

Einstein's search for such a theory during the second half of his life caused him to become isolated from the community of physicists. Most of Einstein's contemporaries felt that the approach was misguided. Einstein, on the other hand, pursued the chimera with such single-minded purpose that he had little time to consider the problems that other scientists thought to be much more important. Einstein was still seeking a unified field theory when he died at the age of seventy-six; at that time he was not much nearer his goal than he had been at the age of forty.

Although Einstein's intuitive feelings about the order of the universe propelled him toward his great achievements, they sometimes brought him into conflict with other physicists, particularly with Niels Bohr. Bohr had convictions about the natural world too. His views and Einstein's did not always coincide.

The arguments between Bohr and Einstein began in 1927 at a conference on physics sponsored by the Belgian industrialist Ernest Solvay. Just a few weeks earlier, Bohr had presented his ideas on complementarity at another conference in Como, Italy. Einstein had not been present at the Como meeting. However, he was in attendance when the Solvay congress opened in Brussels.

The Solvay conference drew many of the most notable physicists of the day. One after another, they gave talks on recent discoveries and theoretical speculations. Einstein did not participate in any of the discussions that followed the lectures. It wasn't until Bohr spoke about complementarity that he rose to take issue.

Einstein's opposition to Bohr's interpretation of quantum mechanics was immediately obvious. In particular, he would not accept the idea that atomic processes were indeterministic. If quantum mechanics could give only statistical predictions, Einstein argued, this meant that it was not a complete theory. When a better, more comprehensive theory was found, the concept of probability would no longer be needed, and determinism would be restored.

Einstein also argued against the acceptance of the Heisenberg uncertainty principle. The idea was logically inconsistent, he claimed. In principle, he said, the position and momentum of a particle *could* be simultaneously determined. Then he gave a theoretical argument in which he attempted to show that this was so.

When Bohr was able to find a flaw in the argument against the uncertainty principle, Einstein was not deterred; he promptly invented a second one. Bohr found a hidden fallacy in that one too. But Einstein still would not give up. His opposition continued long after the Solvay meeting; he continued to dream up new arguments over the course of several years.

By 1930, Einstein was ready to admit that his attempts to prove the inconsistency of the uncertainty principle had been unsuccessful. However, he continued to maintain that the indeterminism of quantum mechanics implied that it could not be a complete theory. "God does not play dice," he would say, over and over again. On one occasion, Bohr became so exasperated that he admonished Einstein, "Stop telling God what to do!"

It is really not surprising that Bohr and Einstein should have argued with each other so vehemently. Their ways of doing physics were too different to permit easy agreement. Einstein was the mystic who could comprehend the universe only in terms of clear-cut pictures which displayed some compelling inner logic. Bohr was the philosopher who was delighted to discover that reality was more profound and paradoxical than it appeared to be. One suspects that Bohr and Einstein did not so much disagree as fail to understand each other. Each was guided by his own inner voice. But the voices did not speak the same language.

Neither outlook on physics can be said to be better than the other. As students of mystical philosophy are fond of pointing out, there is more than one path to the truth. However, the majority of physicists today tend to think that it was Bohr, not Einstein, who was right about quantum mechanics. No deterministic hidden-variable theory of quantum behavior has been found. Or at least, none has been found that seems adequate. The Copenhagen interpretation remains the standard interpretation of quantum mechanics; physicists generally accept the idea that the behavior of subatomic particles can be interpreted only in terms of probabilities.

In 1927, the problem that confronted physicists was one of interpretation. It was already apparent that quantum mechanics would be a successful theory. It contained a mathematical formalism that could be used to perform accurate calculations which had already been verified by experiment. However, the mathematical formulas didn't tell physicists what an electron or a photon "really was." Bohr had developed his ideas on complementarity to fill this gap. The only way to gain a true understanding of quantum reality, he maintained, was to assume that quantum systems could possess contradictory properties. They did not behave like the objects one encountered in the macroscopic world. Concepts that could not simultaneously be applied to ordinary large systems might very well have to be combined if one wanted to understand events in the atomic realm.

The doctrine of complementarity did not represent Bohr's first attempt to introduce apparently contradictory ideas into physics. He had already done this when he proposed a theory of the hydrogen atom in 1913.

Although this theory of Bohr's was eventually superseded by quantum mechanics, it was quite an advance at the time. It was the first successful attempt to apply Planck's ideas about quanta to atomic structure. Bohr's theory did work. It was only when one tried to extend it to atoms that were more complicated than hydrogen that it ran into trouble.

In 1911, the British physicist Ernest Rutherford had discovered the nucleus of the atom. Prior to this time, it had been thought that atoms were tiny balls of positively charged matter in which the negatively charged electrons were embedded. This idea was sometimes referred to as the *plum-pudding model*: the positively charged sphere was the pudding, and the electrons were the plums.

Rutherford discovered that this model was incorrect. He carried out a series of experiments which demonstrated that the positive charge was confined to a tiny sphere in the center of the atom. Presumably, the electrons revolved around this nucleus in the same manner in which planets revolve around the sun.

But there was one very significant difficulty. This planetary model seemed to imply that atoms did not exist. According to the laws of classical physics, the orbiting electrons would radiate energy. They would radiate so much so rapidly that they would quickly lose their energy of motion and fall into the nucleus. The planetary model seemed to contain a serious contradiction.

The problem was solved by Bohr in a very surprising way. He simply assumed that the laws of classical physics did not apply. Electrons in atoms did not continuously radiate away their energy, he said, even though free electrons would do so under similar circumstances.

Furthermore, Bohr suggested, the motion of the electrons was *quantized*. Only certain specific orbits were possible; an electron could not follow any path that lay between them. If one made this assumption, Bohr pointed out, then Planck's quanta of light could easily be explained. An atom emitted a photon when an electron jumped from one orbit to another. The different orbits had different energies. If an electron jumped from a higher energy state to a lower one, the difference was equal to the energy of a particle of light.

Bohr's theory was at least as audacious as any of Einstein's. If this description was accurate, it implied that an electron could vanish and reappear somewhere else. After all, if intermediate orbits were not allowed, that was the only kind of motion between orbits that would be possible.

Bohr's theory embodied several different sets of contradictory ideas. First, Bohr assumed that a free electron could radiate energy, but that electrons in atoms did not (except when they jumped). He assumed that electrons revolved around the nucleus with a smooth, continuous motion, but that the emission of light was associated with motion that was discontinuous. Finally, although a free electron could presumably occupy any position in space, atomic electrons could exist only in a certain set of allowed orbits.

The theory sounded outlandish. However, it won acceptance quickly, for it worked. It explained why hydrogen gas could emit light only at certain specific wavelengths, and it accurately predicted just what those wavelengths should be.

When quantum mechanics superseded Bohr's theory, the idea of electron orbits was dropped. The discovery that the positions of electrons were defined by probability waves made such an interpretation impossible. Bohr's concept of electron jumps was retained, however. Although physicists no longer speak of electron orbits, they still talk about *transitions* from one *quantum state* to another. There are no orbits in the classical sense of the term. However, configurations of probability waves can undergo discontinuous change.

It is not surprising that a physicist who could develop a theory like this should be the one who would later develop an interpretation of quantum mechanics which depended upon the application of contradictory ideas. The principle of complementarity seems to have been a natural outgrowth of Bohr's outlook. Indeed, he was always interested in the application of contradictory ideas, in any field. He was intrigued by philosophical writers who dealt with such notions, and he was interested in the Chinese concept of *yin* and *yang*. Bohr was no mystic, and he was not a devotee of Eastern philosophy. Nevertheless, he placed the yin-yang symbol on the coat of arms that he was required to devise when he was awarded the Danish Order of the Elephant in 1947. Above the symbol, he inserted the legend CONTRARIA SUNT COMPLEMENTA. Contraries are complementary.

If Bohr's outlook on physics was unlike those of Planck and of Einstein, his working methods were different too. While Planck and Einstein worked in solitude, Bohr developed many of his most important ideas in conversation with other physicists. His favorite method of working was thinking aloud. In order to do this, he needed a listener. Bohr's ideas about complementarity were developed, to a large extent, in conversations that he had with such scientists as Heisenberg, Schrödinger and Pauli at his institute in Copenhagen. One suspects that Bohr did most of the talking. A cartoon drawn by the Russian émigré physicist George Gamow shows Bohr with another physicist who has been gagged and bound to a chair. "Please, please," Bohr says, "may I get a word in?"

Bohr, Einstein and Planck had different approaches to the solution of theoretical problems in physics. Sometimes these differences were reflected in their personal lives. Bohr, who was so preoccupied with the role of contradictory ideas in physics, seems to have become interested in the reconciliation of seemingly incompatible ideas while he was still quite young. As an adolescent, he was already pondering such problems as the determinism/free-will dichotomy, coming to the conclusion that the two concepts were really not irreconcilable. In later years, he applied the notion again and again in his work on physics that an idea and its opposite could both be profound truths.

On the other hand, Einstein seems to have looked for solutions that were more clear-cut. In physics, he demanded that a theory be unambiguous and that its meaning be transparent. Like Bohr, he extended this outlook to areas outside of physics. Rather than try to reconcile contraries, as Bohr did, he would opt for one solution or another. For example, after pondering the question of free will, he seems to have decided that it was most logical to conclude that it did not exist. In his book *The World as I See It*, Einstein wrote:

> I do not at all believe in human freedom in the philosophical sense. Everybody acts not only under external compulsion but also in accordance with inner necessity.

Such a point of view is certainly philosophically permissible. However, it is the sort of idea with which Bohr would not have been able to agree.

The differences between Einstein and Planck were dramatic too, although they expressed themselves in different ways. Compared with Einstein and with Bohr, Planck was a conservative physicist. He appears to have been equally conservative in other matters. Planck had great respect for authority, and his manner was reserved and formal. He wore dark clothing, and shirts that were heavily starched. He lived a life that was characterized by its order. He would leave home for his office at the university at exactly the same time every day. A visitor in his household once observed that every morning, just as a clock in the hall was sounding, Planck would emerge from his room and make his way down the stairs to the front door. A part of Planck's day was always reserved for a walk, and he regularly devoted thirty minutes to playing the piano.

The analogy between Planck's life and his work in physics is striking. Unlike Bohr, he did not try to impose his philosophical ideas on physics. His work as a scientist was characterized by a search for order. He seems to have been attracted to the blackbody problem because it appeared to be such a blemish in the structure of theoretical knowledge. It is almost as though he felt that he could not rest until the blemish was removed.

Einstein, on the other hand, was a rebel who never seemed to fit in. As a child, he insisted on going his own way. The very same trait expressed itself again in his later life, when he insisted on laboring away at his unified field theory even though most of his colleagues doubted that he would have any success. If anything, he exhibited an even greater rebelliousness as a young man, when he propounded new ideas that toppled long-hallowed scientific theories.

As a boy, Einstein rebelled against the regimentation that was common in German schools. As a result, he never did manage to graduate from the *Gymnasium* (academic high school); he had already decided to drop out when he was expelled on the ground that "your presence in the class is disruptive and affects the other students."

Einstein's lack of a *Gymnasium* diploma was not the impediment that it might be today. He was able to attend the Swiss Federation Polytechnic School in Zurich, which offered one of the best scientific educations of any university in Europe outside of Germany. But the same problems soon reasserted themselves. Einstein did not get along with his professors particularly well, and this made it difficult for him to get a job after he left the university. Although he graduated in 1900 with an average of 4.91 out of 6.00, nearly two years was to pass before he found permanent employment at the Swiss Patent Office in Bern. It was during his tenure at the Patent Office that he wrote his papers on the quantum theory of light and on the special theory of relativity (and his doctoral dissertation as well).

As an adolescent, Einstein rebelled against authority. As a young man, he began to question the entire structure of physics. It is not surprising that many of the physicists of the day should have looked askance at some of his conclusions, even while they recognized his genius. Even Planck, the ardent supporter of the special theory of relativity, exhibited this kind of attitude at times. When Planck proposed Einstein as a member to the Royal Prussian Academy of Science in 1912, he included a note of apology in his recommendation. Einstein's theory of light quanta, Planck said, "cannot really be held against him." Planck may have been demonstrating his characteristic reserve when he said this. As he must have been aware, other physicists had been branding Einstein's ideas as "reckless."

Although the unconventionality of Einstein's outlook does much to explain the questioning attitude that he exhibited as a scientist, it tells us nothing about where his ideas came from. Nor does Bohr's preoccupation with dichotomies explain where he got the idea for his theory of the hydrogen atom. Planck may have been a scientific conservative. However, there must have been hundreds of conservative scientists among his contemporaries. Most of them have been forgotten, while Planck is remembered as the originator of the quantum theory.

Studies of a scientist's or an artist's personality can sometimes help us to understand his stylistic quirks. But such psychoanalyzing does little to explain the nature of scientific or artistic creation. Although it is clear that the creative imagination plays as important a role in science as it does in art, it is not so easy to say precisely what creativity is. It is clear that the artist and the scientist both make use of intuitive leaps. But it is not so easy to see why scientific intuition such as that possessed by Bohr, or by Planck, or by Einstein should have turned out to be unerringly accurate so often.

We commonly demand that works of art be imaginative. Significant new scientific ideas need to be imaginative as well. In addition, they must possess another quality. They must be correct. However brilliantly imaginative a scientific theory may be, it is worthless if it is wrong.

A correct theory is one that can presumably be verified by experiment. And yet, in some cases, scientific intuition can be so accurate that a theory is convincing even before the relevant experiments are performed. Einstein—and many other physicists as well—remained convinced of the truth of special relativity even when Kaufmann's experiments seemed to discredit it.

How can such things be? Science is supposed to depend upon experimental verification. Scientists habitually remain skeptical of unproved ideas. Can there really be any justification for the acceptance of theoretical ideas that are not backed up by a mass of evidence?

Justified or not, it is something that happens all the time. Since the beginning of the scientific renaissance of the sixteenth century, numerous scientific theories have seemed so plausible that they have won acceptance long before the relevant experiments could be performed. There have been other theories which seemed so implausible that most scientists did not think it worth their while to check them.

Naturally, there have been mistakes. However, scientific intuition has proved to be unerringly accurate on a surprising number of different occasions. Although scientists, like other human beings, often blunder, they have been astonishingly successful in finding significant patterns in the natural world even when they used no tools other than creative thought.

Chapter 4

THE USES AND PITFALLS OF INTUITION

THE ANCIENT BABYLONIANS were superb astronomers. Although they had no telescopes or clocks, or any of the other instruments that can be found in a modern observatory, some of their observations had an accuracy that was not equaled until the nineteenth century. The Babylonians were infallible in their forecasts of lunar eclipses; they could even predict how complete an eclipse of the moon would be. Some of the calculations performed by the Babylonian astronomer Kidinnu in the fourth century B.C. have turned out to be more accurate than those made by European astronomers as late as 1887.

But the Babylonians had no theory to explain why the bodies in the solar system moved the way they did. They could predict eclipses and conjunctions of planets, but they had no idea what a planet was. It wasn't until astronomy was taken up by the ancient Greeks that the first attempts were made to find theoretical explanations for the motions that were observed in the heavens.

By modern standards, the Greek theories were not very good ones. Although the Greeks made an important advance when they realized that celestial events must have logical explanations, they relied a little too much on principles they thought to be intuitively obvious. As it turned out, some of these "obvious" principles were wrong.

According to the philosopher Plato, it was only reasonable to believe that the motions of the celestial bodies were as perfect as their Maker could frame them. To Plato, this meant that the planets had to move in circles at uniform velocity. Nothing else was even conceivable; the circle was the most perfect geometrical figure.

There was one serious problem with this view. As seen from the earth, the motions of the planets hardly seem regular at all. They do not appear to move in circular paths, as the stars do. At certain points in their orbits, planets will suddenly change their direction and move backward with respect to the fixed stars. The apparent irregularity of their behavior is reflected in the etymology of the word. "Planet" derives from a Greek word meaning "wanderer."

The tendency of the planets to retrogress, or move backward, can be explained quite easily if one only assumes that they revolve around the sun, not around the earth. But this was an assumption that Plato and his followers were not willing to make. They believed that the most significant difference between the earth and the heavens was that celestial bodies were immutable. Only on the earthly sphere were there birth and death, growth, change and destruction. To imagine that the earth was a planet that moved around the sun seemed contradictory. To Plato, it was absurd to believe that the earth could be a part of the perfect and changeless heavens.

There seemed to be other arguments for believing in an earth-centered universe as well. If the earth moved, it was argued, then the stars would seem to shift their positions as it traveled in its orbit. The constellations would not appear the same in summer as they did in winter.

In fact nearby stars do exhibit a shift when they are viewed from opposite ends of the earth's orbit. Their apparent positions in summer and in winter are not precisely the same. However, this *stellar parallax* cannot be observed by the naked eye. Plato could not have known that it was real, since he lived two millennia before the invention of the astronomical telescope.

Another argument against a moving earth had to do with the consequences that the ancient Greeks thought would follow from the earth's rotation. They believed that if the earth rotated, then an object that was thrown up into the air would not come back down in the same spot. The earth would rotate under it, and it would be left behind. Birds would have to fly rapidly just to keep up with the moving earth, while the air and clouds would be left behind.

Some Greek philosophers, for example the followers of Pythagoras in the fifth century B.C. and Aristarchus of Samos in the third century B.C., did hypothesize that a rotating earth moved in a circular orbit around the sun. However, their views were rejected by most of the astronomers and philosophers of the ancient world. Such a view was held to be an affront to common sense. There were too many arguments which seemed to indicate that it was the heavens that moved.

The first attempt to explain why the planets move about in such an irregular way was made by Plato's pupil Eudoxus during the fourth century B.C. According to Eudoxus, the heavens were made up of a system of transparent, interconnected spheres. The outmost spheres carried the stars around the earth. Complicated systems of inner spheres accounted for the motions of the planets. The planet Jupiter, for example, was embedded in a sphere that interacted with three others. Its apparently irregular path in the sky was a combination of four different circular motions.

The system was improved upon by Aristotle, who increased the total number of spheres from twenty-six (six for the sun and moon, and four for each of the planets known to the Greeks) to fifty-five. But Aristotle's system was incapable of explaining all the observed facts. If the universe was made up of geocentric spheres, then it seemed to follow that the planets must always remain at the same distance from the earth. But if the distances did not change, it was hard to understand why the planets should vary in brightness. The existence of brightness variations seemed to imply that they were sometimes closer to the earth, sometimes farther away.

So Greek astronomers and mathematicians developed a new theory to explain planetary motion. They discarded the concept of concentric spheres and replaced it with a theory of *deferents* and *epicycles*. A deferent was an earth-centered circle. An epicycle was a smaller circle whose center was on the deferent. An epicycle can be thought of as a small wheel rolling around on a larger one.

If a planet was attached not to the deferent, but to the rim of a moving epicycle, its motion—as seen from the earth—would not appear to be circular. The planet would go through a series of loops. Since it would sometimes loop backward for short periods, the retrogressions were thus explained.

By using systems of epicycles and deferents, the Greek astronomers managed to avoid contradicting Plato's dictum that only uniform circular motion was possible. However, the epicycle/deferent theory worked in only an approximate way. Planetary motions calculated with the theory did not correspond exactly to observed motions.

Consequently, additional mathematical devices were introduced. Epicycles were placed on epicycles, and concepts such as the *eccentric* and the *equant* were added to the system. The eccentric was a device for displacing the centers of the planetary circles from the earth to some point in space. The equant was a mathematical concept introduced to explain the observed nonuniform motion of the planets. If a planet did not always move through the sky at the same velocity, the Greek astronomers assumed that its motion was uniform with respect to a point that was displaced from the center of the circle. It was a subterfuge used to permit motions to be described as "uniform" when in fact they were not.

By the time of the Alexandrian astronomer Ptolemy (second century of the Christian era), astronomical theory had become complicated indeed. Not only were there epicycles on epicycles, one also had to use eccentrics on deferents and eccentrics on eccentrics to reconcile theory and observation. The system had become so complicated that astronomers no longer pretended that it had anything to do with reality. They believed that it was their task to "save the appearances," by finding a way to reduce planetary motions to circles.

Ptolemy was quite explicit about this. "We believe that the object which the astronomer must strive to achieve is this: to demonstrate that all phenomena in the sky are produced by uniform and circular motions," he wrote. Only such motions, Ptolemy claimed, were appropriate to the planets' divine nature. But it didn't make any difference what mathematical construction one used as long as it gave the correct result. If a particular combination of epicycles gave the same answer as a combination of eccentrics, the astronomer was free to use whichever technique he desired.

In Ptolemy's time, it was thought that physics was irrelevant to astronomy. Mathematics and theology were the disciplines appropriate for understanding the events that took place in the heavens. An astronomer should not ask whether planets could really move that way. On the contrary, he should study the abstract mathematical techniques in order to establish harmony in his soul.

Ptolemy described his system in a book whose Greek title can be translated as *Mathematical Concordance of Astronomy*. During medieval and Renaissance times, this work was known as the *Algamest*. *Algamest* was the Latin version of an Arabic corruption of a Greek phrase which meant "the greatest." The title was not an inappropriate one, for the *Algamest* was to represent the last word in astronomy for well over a thousand years.

Satisfaction with the Ptolemaic system was apparently not universal. For example, during the thirteenth century King Alfonso X of Castile remarked that if the Almighty had consulted him before embarking on the Creation, he could have recommended something simpler. Nevertheless, it wasn't until the sixteenth century that Ptolemy's theory was seriously questioned.

In 1543, the Polish canon Niklas Koppernigk, better known by his Latin name Copernicus, published a work called *On the Revolutions of the Heavenly Spheres.* Copernicus had delayed publishing the book for thirty years. The first completed copy arrived from the printer only a few hours before his death.

No one really knows why Copernicus put off publication for so long. It has been suggested that he feared ecclesiastical censure, that he may have feared the ridicule of other astronomers and even that he was afraid that his theory was not good enough. But whatever the reasons for the delay, the treatise did finally make its way into print. In the view of many historians, its publication marked the beginning of a scientific revolution that has continued to this day.

As everyone knows, Copernicus argued that the planets revolved not around the earth, but around the sun. It is not so widely known that Copernicus was not the first to advance this hypothesis, or that his system was every bit as complicated as the Ptolemaic one.

Copernicus' achievement was not the invention of a sun-centered solar system. As we have seen, this idea had already been suggested during ancient times. Furthermore, the revival of learning during the Middle Ages had led to speculation on this subject long before Copernicus was born. Arguments in support of the idea that the earth moved had been given by medieval scholars nearly two hundred years before Copernicus' book was written.

Nevertheless, it is Copernicus who is credited with the theory of the heliocentric solar system. This is not inappropriate, for it was Copernicus who transformed vague speculations about a moving earth into a consistent theory. Not only did he give arguments for the rotation of the earth and its revolution around the sun, he also developed mathematical constructions that would allow one to compute the motions of the planets. Copernicus' work was more than speculation. His theory could be used to construct tables that described planetary movements as they were seen from the earth.

Copernicus gave no proofs that the earth moved around the sun. In his day, none existed. In fact, to an unprejudiced observer it must have seemed that the arguments for a stationary earth were stronger. There was as yet no way to detect stellar parallax, and since the principle of inertia was not yet understood, there was no good way to refute the old Greek argument that a moving earth would leave clouds, air and birds behind.

Copernicus did not attack Ptolemy's theory because he thought that the arguments for a rotating earth were better. His objections were of a purely esthetic nature. The Ptolemaic system, he charged, was "inconsistent and unsystematic." It was simply too complicated. Copernicus could not believe that such a complex system of epicycles and deferents, of eccentrics and equants, was an accurate representation of reality.

As we have seen, this was more or less what Ptolemy had believed. The difference between them was that Ptolemy had considered astronomy to be a branch of theology, while Copernicus wanted to make it part of physics. Ptolemy had been content with "saving the appearances." Copernicus wanted a theory that was intuitively reasonable.

When he worked out his system, Copernicus retained the old idea that the orbits of the planets were composed of uniform circular motions. He had no way of knowing that planetary orbits were elliptical rather than circular. Nor was he aware that the planets moved faster when they were nearer the sun and more slowly when they were farther away. Hence, like Ptolemy, he was forced to resort to the use of epicycles to explain the irregularities in their motion. At first, he thought that thirty-four circles would suffice. In the end, he wound up using nearly fifty. Copernicus was not able to dispense with eccentrics either. In the Copernican system, the earth did not revolve precisely around the sun, but rather about a point in space that was near it.

As a tool for calculation, the Copernican system was no worse than the Ptolemaic. But neither was it any better. As a result, the astronomers of the day ignored Copernicus' principles, and used his methods whenever they happened to be convenient. They complacently used the Ptolemaic system as long as it gave accurate results, switching to the Copernican theory in those cases where it seemed to work better. Few of them took Copernicus' idea of a sun-centered system seriously. In their eyes, it was a false hypothesis which could serve as a useful tool for calculation.

This attitude was perhaps best summed up by Copernicus' colleague Osiander. Entrusted with the task of overseeing the printing of *On the Revolutions*, Osiander added a preface of his own without bothering to obtain Copernicus' permission. In it, he gave the following admonishment to the reader:

> Let us therefore permit these new hypotheses to become known together with the ancient hypotheses, which are no more probable; let us do so especially because the new hypotheses are admirable and also simple, and bring with them a huge treasure of very skilful observations. So far as hypotheses are concerned, let no one expect anything certain from astronomy, which cannot furnish it, lest he accept as the truth ideas conceived for another purpose, and depart from this study a greater fool than when he entered it.

It was a sentiment which even Ptolemy could have accepted, and one with which most of Copernicus' readers tended to agree.

It has been said that revolutionary new scientific ideas eventually triumph because their opponents grow old and die and are replaced by new generations of scientists who grow up with the new notions. This is probably no longer true today; if it was, science could not advance as rapidly as it does. In Copernicus' time, however, scientific revolutions happened more slowly.

The new heliocentric theory was not accepted by Copernicus' contemporaries. By the end of the sixteenth century, however, there existed a new generation of scientists. Some of the members of this generation, notably Galileo and the German astronomer Johannes Kepler, viewed matters in an entirely different light.

While he was still a student at the University of Tübingen, Kepler heard about Copernicus from Michael Maestlin, his professor of astronomy. Maestlin was one of the early converts to the Copernican theory. He had no difficulty converting Kepler, who readily agreed that the only reasonable planetary arrangement was one which was centered around the sun. Throughout his life, Kepler never wavered from this belief. Nevertheless, before he could elaborate upon it and make contributions of his own to astronomy, he had to go through some very curious detours.

Kepler published his first book, *Mysterium Cosmographicum*, in 1596, when he was twenty-five. In this book, he took the Copernican system for granted, and attempted to add to it by describing a discovery of his own. The "discovery" would eventually prove completely wrong. Nevertheless, Kepler would spend decades elaborating upon it. During most of his life, he continued to believe it the most important discovery he had made.

The insight that led to this intriguing but mistaken idea happened in a classic manner. While drawing a figure on the blackboard during a lecture on astronomy in the Austrian provincial capital Graz in 1595, Kepler was suddenly struck by an idea. "The delight that I took in my discovery," he was to write later, "I shall never be able to describe in words." The idea was to be the inspiration around which Kepler's life revolved for years.

Kepler's idea came only after years of thought about the Copernican system. When still a student of astronomy, Kepler had begun to wonder why there should be only six planets (these were Mercury, Venus, the earth, Mars, Jupiter and Saturn; Uranus, Neptune and Pluto would not be discovered until much later), and why they followed the particular orbits that they did.

At first, Kepler wondered whether one planetary orbit might not be two, three or four times as large as another. But when he tried these and other simple numerical proportions, he could find none that worked. Since Kepler could not believe that the arrangement of the orbits was devoid of order, he continued to work on the problem. Years went by, and he had no success.

Today we know that there are no simple numerical relationships between the sizes of the various planetary orbits,* and there seems to be no good theoretical reason why there should be. But

---

* There is a relationship, called *Bode's law*, which is approximately correct for the planets out to Uranus, provided that one treats the asteroid belt between Mars and Jupiter as a "planet" also. However, the "law" fails dismally in the cases of Neptune and Pluto. Its partial success is believed to be nothing but a numerical coincidence.

to Kepler, it was intuitively obvious that some sort of relationship had to exist. He believed that the solar system was the handiwork of God. It was inconceivable that the Creator should have constructed the solar system without some definite plan.

While Kepler was drawing a geometrical diagram on the blackboard in 1595, he was suddenly struck by the fact that the relationships between the orbits might be of a geometrical, rather than a numerical, nature. Once he had the idea, the rest was relatively simple. It wasn't long before Kepler had worked out a theory.

Kepler's theory was based on a discovery that had been made by the ancient Greek mathematician Euclid. Euclid had proved that there existed only five "perfect" geometrical solids. A perfect solid is a three-dimensional figure which possesses faces that are perfectly symmetrical.

The simplest of the five solids is the tetrahedron, a three-sided pyramid that comprises four equilateral triangles (the fourth triangle is the base of the pyramid). The next is the cube, which consists of six squares. The remaining three solids are the octahedron (eight equilateral triangles), the dodecahedron (twelve pentagons) and the icosahedron (twenty equilateral triangles). No one has ever found a way to construct a sixth perfect solid, and no one ever will. Euclid's theorem proved that such a construction is impossible.

Noting that there were five solids and six planets, Kepler wondered whether the solids might not have something to do with the spacing between the orbits. He quickly tried out a number of different schemes. Soon he had one that seemed to work.

When Kepler placed a cube inside the orbit of Saturn, arranging it so that the corners of the cube just touched Saturn's "sphere," he found that the orbit of Jupiter would just fit inside the cube. Next, he placed a three-sided pyramid—the tetrahedron —inside Jupiter's orbit, and found that the space inside the pyramid was just large enough to contain the orbit of Mars. Using the dodecahedron next, and then the icosahedron and octahedron, he obtained the orbits of the earth, Venus and Mercury.

Kepler thought he had discovered the secret of the cosmos. It was a delusion. The only reason that his geometrical theory seemed to work was that the astronomical observations which existed at the time were full of errors. Thus whenever Kepler failed to obtain an exact fit, he could blame the discrepancy on flaws in the available astronomical data.

In a way it is fortunate that Kepler pursued his delusion for so long. Without such a "discovery" to motivate him, he might not have found the perseverance that he was to need in order to find the true laws of planetary motion. Kepler had wandered into a blind alley. But his exhilaration over what he thought he had found there carried him through decades of tedious mathematical labor, and eventually allowed him to gain scientific immortality.

As the years passed, Kepler began to have problems with his geometrical theory. As more accurate data became available, he began to realize that the regular solids did not fit perfectly between the planetary orbits. In the end, Kepler was to realize that the planetary orbits were not circular at all. He would discover that he was forced to give up his vision of geometrical harmonies in favor of what he described as a "cartful of dung": elliptical orbits.

Planetary orbits are only approximately circular. They are slightly elongated; the distance from a planet to the sun is not the same at all times of year. For example, the average distance from the earth to the sun is 149.6 million kilometers (about 93 million miles). But it is only 147.1 million kilometers when the earth is closest, and 152.1 million kilometers when it is farthest away. The paths of some of the other planets are even more elongated. Mars, for example, approaches to within 206.7 million kilometers of the sun, and recedes to 249.1 million.

If Kepler had not been driven by his illusions, he might never have discovered these facts. However, motivation was not the only thing that was needed. If Kepler had not eventually been able to gain access to accurate astronomical observations, there would have been no need to give up the geometrical theory. Kepler would never have made the discoveries for which he is remembered today if the Danish nobleman Tycho Brahe had not conceived a passion for making meticulous astronomical observations.

Brahe was one of the most extravagant characters in the history of science. Granted a 2,000-acre island by the Danish king, Frederick II, Tycho ruled his small domain like a medieval tyrant. He was rude to all those who displeased him, including King Christian IV, Frederick's successor. He extracted labor and goods to which he was not entitled from his tenants, and flouted decisions of the provincial Danish courts, and the High Court of Justice as well, by leaving a tenant and all his family in chains.

Brahe had no flashes of insight like those which Kepler experienced. However, without Brahe—or an astronomer like him—a correct theory of planetary motion could not have been conceived.

The observatory that Tycho built for himself on his island looked like a fortress. According to author Arthur Koestler, it resembled "a cross between the Palazzo Vecchio and the Kremlin." The observatory possessed galleries that were filled with clocks, sundials, globes and allegorical figures. A printing press and an alchemical furnace shared space in the cellar with Tycho's private dungeons. The estate had game preserves and artificial fishponds, and Tycho had his own pharmacy and paper mill as well.

Tycho's astronomical instruments were elaborate indeed. Some of them cost more money than Kepler made in his lifetime. But Tycho did not confine his extravagances to astronomy. Life at his observatory was enlivened by a steady succession of banquets, which were presided over by the hard-drinking master, who could often be seen throwing morsels of food to his court dwarf or rubbing ointment on his silver nose.

Tycho had lost the bridge of his nose in a duel that took place while he was still a student. The duel was supposedly the result of an argument over who was the better mathematician. Tycho replaced the missing part of his nose with an alloy of gold and silver. It must have given him an outrageous appearance that was strangely in keeping with his character.

Tycho died in 1601, of a burst bladder. It seems that he had been drinking heavily at a dinner given by the Baron Rosenberg in Prague, and had not gotten up to urinate. During the last night of his life, while experiencing mild delirium, he repeated over and over, "Let me not seem to have lived in vain."

Kepler, who had been working as Tycho's assistant at the time of his death, saw to it that he did not. He accomplished this by stealing Tycho's astronomical observations. Kepler was afraid that if they fell into the hands of the heirs, he might never be able to gain access to them. Kepler was quite candid about the theft. Just a few years later, he described the event to an English admirer in the following words:

> I confess that when Tycho died, I quickly took advantage of the absence, or lack of circumspection, of the heirs, by taking the observations under my care, or perhaps usurping them.

Once he was in possession of the data, Kepler began to work out the orbits of the planets in detail. The orbit that gave him the greatest difficulty was that of Mars. However he struggled, Kepler could not find a way to fit the data to a circular orbit. The best that he was able to do was match theory and observation to an accuracy of eight minutes of arc (two-fifteenths of a degree). He tried one circular orbit after another. Whatever he did, the annoying discrepancy remained.

If Kepler had been working with observations made by anyone but Tycho, he could have disregarded the difference. The data collected by Tycho's predecessors had frequently contained errors that were even larger. But Kepler knew that Tycho had been the most meticulous observer in the history of astronomy; the eight-minute difference could not be passed off as a result of observational inaccuracy.

Kepler struggled with the problem for years. In the end, he was forced to give up the idea that the planetary orbits were circular. It was an ellipse that described the orbit of Mars. Kepler did not know why this should be, and he was very much aware that the concept of elliptical orbits was in conflict with his vision of universal geometrical harmony. He knew, however, that no other conclusion could be reached.

After Kepler had determined that the orbits of the other planets were elliptical also, he formulated two other laws of planetary motion. One of these described how the velocity of a planet changed at different points in its orbit. The third related the distance of a planet from the sun to the time that it took to complete a revolution. The three laws must be considered to be Kepler's greatest achievement. Without them, Newton would not have been able to derive his law of gravitation.

When Kepler announced the third of his laws in 1619, Newton had not yet been born. No one, not even Kepler himself, understood the significance of the laws of planetary motion. At the time, there were numerous astronomers who would not even consider the possibility that the sun might be the center of the planetary system. Furthermore, a certain amount of religious opposition to the heliocentric idea had developed. Once Roman Catholic and Protestant theologians realized what scientists like Copernicus and Kepler had been doing, they were quick to point out that the Bible seemed to imply that it was the sun that moved around the earth.

The scientist who did the most to gain acceptance for the Copernican ideas was Galileo. Galileo, however, would not accept Kepler's laws. He refused to believe that planetary orbits could be elliptical, and advocated an astronomical system that was still based on circles.

There are so many myths concerning Galileo that it might be best to attempt to explode a few of them before we go on. To begin with, Galileo never dropped two balls—a light one and a heavy one—from the Leaning Tower of Pisa in order to demonstrate that they would fall at the same rate. This experiment was actually carried out by one of Galileo's scientific opponents, who wanted to show that the heavy ball would strike the ground first. In fact, this is exactly what happened. The light ball lagged behind because it was impeded by air resistance. Light and heavy bodies fall at the same rate only in a vacuum, or when the effects of air resistance are so small that they can be neglected.

Galileo was not quite the martyr to science that he is sometimes thought to have been. The Roman Catholic Church did force him to recant his belief in the Copernican system of astronomy. But he was never put in prison. When the Church held him in detention while preparing his trial, Galileo was lodged in luxurious apartments which overlooked St. Peter's and the Vatican gardens. Galileo was threatened with torture, but at the same time he was treated with a certain respect. After all, he was one of the best-known scholars of the day.

Galileo seems to have brought the confrontation with the Church on himself. Church officials had previously advised him that he could teach and expound the Copernican theory as a hypothesis, and as a tool that was useful for doing astronomical calculations, as long as he did not propound it as a true theory.

And of course this is exactly what Galileo did. He tried to be devious about it. But his subterfuge was so easy to see through that the Church officials must have believed that Galileo took them to be fools. In his book *Dialogue on the Great World Systems*, which was published in 1632, Galileo invented three characters who gave arguments for and against the Copernican hypothesis. He put the arguments against Copernicanism into the mouth of one called Simplicio, who has been described by Koestler as a "good-natured simpleton." It was hardly possible to read the book and believe that the author considered the idea of an earth-centered universe to be valid. It is not surprising that Galileo should have had difficulties with the Church.

Galileo's *Dialogue* is a curious book. It was written because Galileo believed that he had convincing evidence for the validity of the Copernican system. Great emphasis is placed on certain ideas that are supposed to clinch the argument. But these ideas prove nothing at all, for they are demonstrably false.

The last section of the book is devoted to Galileo's theory of the tides. Not only is this theory incorrect; it is based on assumptions which, as even Galileo was aware, are contradicted by observed facts. Furthermore, when Galileo propounded this theory, he had to ignore an existing hypothesis that was completely correct.

Kepler had surmised correctly that the tides are caused by the gravitational attraction of the moon. Although decades were to pass before Newton would formulate a law of gravitation, Kepler seems to have rather vaguely understood that the moon could exert some kind of pull on the waters of the earth's oceans. Now, tidal attraction has nothing to do with the correctness of Copernicus' heliocentric theory. The moon revolves around the earth in both the Copernican and Ptolemaic systems. The causes of the tides are not relevant to the question of which theory is correct.

However, Galileo was so anxious to find a way to substantiate the Copernican hypothesis that he invented an explanation of his own, formulating it in such a way that it would seem to provide evidence for the heliocentric system. Dismissing Kepler's theory as astrological nonsense, Galileo proclaimed that the earth's tides were a consequence of its motion. In his theory, the tides were caused by a kind of sloshing back and forth of the water in the oceans. This sloshing was supposedly a result of the earth's rotation and its movement around the sun.

This tidal theory contradicted Galileo's own discoveries concerning the laws of motion. Furthermore, it did not agree with known facts. When it was worked out in detail, Galileo's theory implied that there should be one high tide per day, and that it should always take place at noon. In reality, there are two high tides in any given twenty-four-hour period, and they occur at different times during different parts of the month. Galileo was perfectly aware of this discrepancy. He attempted to explain it away as the consequence of certain other causes, such as the shape and depth of the sea.

If Galileo's reasoning was too contrived, at least his motivation for propounding the tidal theory is understandable. Galileo *knew* that the heliocentric theory was correct, but he did not know of a way that he could convincingly demonstrate it. So when all else failed, he resorted to fantasy.

Scholars agree that the Church's suppression of Galileo's book impeded scientific progress. However, there was some justification for the position taken by the Church. Galileo did not possess any scientific evidence to demonstrate that the earth moved. He was able to present arguments for the hypothesis. However, as he realized very well, it was possible to argue against the theory also.

It is not my intent to belittle Galileo's achievements. He was the greatest scientist of the seventeenth century. His investigations of moving bodies provided the foundation upon which Newton's laws of motion would be based. Galileo was the first to understand the concept of acceleration, and he singlehandedly discredited much of Aristotelian physics. His contributions to astronomy were equally important. He was the first to study the heavens with a telescope. With this instrument, he observed sunspots, showed that the planet Venus had phases like those of the moon and discovered four of the moons of Jupiter.

None of these discoveries, however, really confirmed the Copernican hypothesis. The fact that Jupiter had moons did not demonstrate that the planet itself did not revolve around the earth. It is true that the Copernican theory predicted that Venus should exhibit phases. But the argument didn't work in reverse: the observation of phases did not demonstrate that Venus revolved around the sun.

Galileo must have been aware that the evidence for a moving earth was, as yet, inconclusive. Otherwise he would not have introduced the tidal theory at the end of his *Dialogue* in an effort to make the case more convincing. Galileo really had no convincing proof, but he was a firm believer in the heliocentric theory nevertheless. Like Copernicus, he did not believe that a theory as complicated and inelegant as the Ptolemaic could possibly be an accurate description of nature.

That sounds a little like intellectual dishonesty. But it is not. It is a sign that Galileo possessed the penetrating intuition that is the mark of a great scientist. He could peer behind appearances and see that the Copernican theory was true, even though he had no conclusive evidence. He believed in it so firmly that he invited trouble from the Church in order to convince his fellow scientists.

By the time that Isaac Newton published his law of gravitation in 1687, the Copernican theory was widely accepted. One by one, the astronomers who had clung to Ptolemaic astronomy had died. Meanwhile, the younger astronomers readily accepted the new ideas. When Newton propounded his theory, controversy did flare up again. But now the argument was about the concept of gravitational attraction, not about the heliocentric hypothesis. Although the Copernican theory had still not been proved, few were willing to question its validity.

The proof that the earth really did travel around the sun was obtained only in 1729, two years after the death of Newton, one hundred and eighty-six years after Copernicus' epoch-making book was published.

In 1725, the British astronomers Samuel Molyneux and James Bradley set up a specially designed telescope to test a claim that had been made by the physicist Robert Hooke. Hooke was a quarrelsome English scientist who had engaged in controversies with numerous individuals, including Newton. It seems that Hooke claimed that he had measured the annual parallax of a star that passed directly over London. As we saw earlier, the absence of parallax was used by Plato as an argument against a moving earth. So if Hooke's assertion was correct, then he had obtained proof that the earth revolved around the sun. But Hooke's observations had been few, and his contemporaries found his results unconvincing. Therefore Molyneux and Bradley decided to repeat the experiment using more accurate instruments.

The two astronomers found that the star did not shift its position in the way that theory predicted. It was apparent that the Copernican hypothesis, although it was universally accepted by now, still had not been proved. As it turned out, the existence of stellar parallax was not to be demonstrated until 1838.

Molyneux and Bradley had observed a shift, but it was not the kind they had been looking for. They did not know how to explain what they had observed. Several years passed. Then, one day, Bradley was suddenly struck by an explanation for the peculiar behavior of the star.

In 1729, Bradley showed that the shifts in the apparent position of the star had been caused by the motion of the earth. He pointed out that if a telescope was moving, stars would appear to occupy slightly different positions than they would if it was stationary. Since the earth moved in opposite directions when it was on opposite sides of the sun, the shift caused by this *aberration of starlight* would be different during the various seasons of the year. Although he and Molyneux had been unable to confirm Hooke's claim to have observed parallax, their experiment had demonstrated that the earth moved around the sun.

The experiment, however, said nothing about the earth's rotation on its axis. Since the earth rotates at a constant velocity, and always in the same direction, the aberration of starlight cannot be used to detect this motion. The rotation does produce a shift, but the shift is always the same, and there is nothing to which it can be compared.

Proof of the rotation of the earth was not obtained until the middle of the nineteenth century. In 1851, the French physicist Jean Bernard Léon Foucault began a series of experiments with pendulums. He was able to show that a large, heavy pendulum would change its direction of motion as the rotating earth twisted under it. After three centuries, Copernicus' assertion that the earth rotated had finally been proved.

There is something very striking about these experimental verifications. By the time they were performed, no one seriously doubted that the earth revolved around the sun or that it rotated on its axis. The experiments did nothing but verify ideas which scientists already found convincing.

Scientists were convinced by Copernicus' and Kepler's and Newton's insights into the workings of the universe. These insights seemed so compelling that they could hardly be doubted. Kepler's laws and Newton's theory had revealed a mechanism which seemed so coherent and so logical that it was hard to imagine that there could be any serious defects in the picture. When the experimental verifications were finally obtained, science gained no new knowledge; it simply tied up some loose ends. It would not be inaccurate to say that the scientific revolution that was begun by Copernicus and completed by Newton was a demonstration of the primacy of theory over experiment.

Copernicus, Kepler and Galileo understood what the celestial motions were like centuries before the appropriate experiments could be performed. It was their insight that led them to their discoveries. And yet this insight misled them as often as it steered them in the right direction. Copernicus fell back on epicycles when he discovered that circles didn't work. Kepler dreamed up a fantastic hypothesis about perfect solids and celestial spheres. Galileo's explanation of the tides can be described only as a crackpot theory. It is clear that scientific insight can lead to error as easily as it can show the way to the truth.

Sometimes false insights are accompanied by a conviction that a great truth has been discovered. Sometimes the conviction is even stronger than those which true discoveries evoke. Kepler, for example, was so overwhelmed by his idea about geometrical harmonies that he continued to elaborate upon the concept for most of his life. By comparison, the three laws of planetary motion were given much less emphasis. They were buried in Kepler's writings, and might have gone unnoticed if Newton had not been perspicacious enough to recognize their importance.

It was the striking and logical nature of their insights that won acceptance for the theories of Copernicus and of Newton, and for the theories of such modern scientists as Einstein as well. But if insight can so often be wrong, why is science not more affected by its pitfalls? Why does the conviction that accompanies false insight not cause science to follow erroneous paths?

The answer, of course, is that this is exactly what happens at times. The intuitive convictions of certain creative individuals have caused errors that have persisted for centuries. Plato's idea that only circular planetary orbits were possible seemed so self-evident for so long that even Galileo could not give it up. Einstein introduced his cosmological constant because he could not give up the centuries-old idea that the universe must be static and unchanging. When he discovered that his theory implied that the universe had to be either expanding or contracting, he refused to believe it, and introduced a fictional force to do away with the "problem."

Histories of science generally deal only with successes. As a result, they tend to ignore mistaken insights and the problems they cause. Consequently, they frequently paint an inaccurate picture of scientific progress. They give the impression that science is made up of a string of successes, when in reality even great scientists make as many mistakes as anyone else.

The progress of science follows an irregular path. Scientists blunder into one blind alley after another. Sometimes they manage to find their way out of them, and sometimes they do not. The mistaken theories that they propose are eventually discarded. But sometimes this happens only when future generations realize that they really do not explain very much. The idea of circular planetary motion was thrown out when it was realized that the concept only created difficulties. Two thousand years passed before Kepler discarded Plato's idea.

If such things can happen, then it is not so obvious how we can be sure that what we take to be science is anything but an accumulation of errors. But perhaps it would be better if I took up this problem in a subsequent chapter. It would be best if we first took a closer look at the roles played by experiment and by theory.

Chapter **5**

THE PRIMACY
OF THEORY

ONE OF THE BEST EXAMPLES of conflict between observation and theory is provided by the story of the gradual acceptance of the theory of continental drift. The idea that continents can move across the face of the earth is one that has been suggested many times. Anyone who looks at a map of the world cannot help being struck by the fact that the South American and African continents look like two pieces of a jigsaw puzzle. It looks as though the east coast of Brazil would fit into Africa's inward curve perfectly.

The striking fit between the two continents was noted as long ago as 1620 by Francis Bacon. Although Bacon failed to draw the conclusion that the two continents were once connected, scientists were making precisely that suggestion by the middle of the nineteenth century. In 1857, the English zoologist Richard Owen speculated that the Atlantic Ocean had been created when the two continents had drawn apart. A year later, in 1858, the American author Antonio Snider proposed a similar theory, suggesting that the Atlantic might have been created during the Biblical Deluge. So much rain had poured down during the Flood, Snider said, that it could have caused a primeval land mass to crack, and to separate into a number of different pieces.

Snider's catastrophic theory was not taken very seriously by professional geologists. Nevertheless, some of them speculated that there might exist natural forces which could cause continents to split and to gradually drift apart. In 1885, the Austrian geologist Eduard Suess published a map that showed how the southern continents could be fitted together into a single supercontinent, which Suess named Gondwanaland after India's Gondwana province. In 1908, the American geologist Frank B. Taylor suggested that continental drift might explain the origin of mountains. Continental movements, he pointed out, would create forces that would compress sections of the earth's crust, causing mountains to rise.

Although Owen, Snider, Suess and Taylor all suggested that it was possible that continents could drift apart, it is the German meteorologist Alfred Wegener who is generally credited with being the originator of the modern theory of continental drift. Where his predecessors generally confined themselves to making suggestions about the possible validity of the idea, it was Wegener who spent years amassing evidence in an attempt to demonstrate that continental drift was a reality.

When Wegener published his book *The Origin of Continents and Oceans* in 1915, it had long been known that there were striking similarities between fossils found on widely separated continents. In some cases the fossils seemed to be identical. For example, fossils of *Mesosaurus*, a reptile that lived near the end of the Paleozoic era, about 270 million years ago, were found in Brazil and in South Africa, and nowhere else.

Wegener added to these observations by noting that there were similarities between living species as well. The lemur, for example, was found in both India and Africa. The garden snail *Helix pomatia* also lived on continents that were separated by oceans; it was found in western Europe and North America alike.

Wegener was aware that continental drift was not the only possible explanation for the distribution of living species. A number of others were frequently cited by biologists. Charles Darwin had suggested in *On the Origin of Species* that the snail might have been carried across the Atlantic Ocean on the feet of birds. Wegener did not consider this idea to be very plausible, however. In any case, it didn't explain how larger animals had migrated. Nor was Wegener impressed by the theory, proposed by many biologists, that there had once existed land bridges between the continents.

The trouble with land bridges was that there seemed to be no way to explain their subsidence. Wegener did not believe that such bridges could have disappeared without a trace. After all, the granite that made up the continental crust was lighter than the basaltic rock of the ocean floors. If land bridges had come into existence and then disappeared, a lighter material would have had to sink into a heavier one.

In fact, the sinking of land bridges contradicted the principle of *isostasy*, which had been formulated by Suess around the end of the nineteenth century. According to Suess, the continents floated on the heavier material that lay beneath them; in effect, the continents were giant rafts.

To provide evidence in support of this idea, Suess had pointed out that it had been known for some time that both Canada and Scandinavia were gradually rising at about the rate of one centimeter per year. According to Suess, this could be explained by the fact that both land masses had been covered with ice during the ice age that had ended 11,000 years ago. When the ice melted, the weight that had been holding Canada and Scandinavia down was removed, and they began to rise. The process was analogous to the rising of a ship in the water when its cargo was unloaded. The only difference was that since the rock which supported continents flowed more slowly, the rise would be much more gradual.

Wegener realized that the principle of isostasy not only made the sinking of land bridges improbable, it also provided an explanation for the similarities between the fossils and the plants and animals that were found in widely separated locations. If the material in the interior of the earth was fluid enough to allow continents to rise, however slowly, then there was no reason why there should not be horizontal movement also. If continents drifted, Wegener realized, plants and animals would not have had to migrate over wide expanses of ocean.

The fit between opposite coastlines, similarities in fossils and living organisms, and the theory of isostasy made the idea of continental drift seem plausible. But they did not provide convincing evidence for the idea. Consequently, Wegener set to work to see if he could find some other kind of evidence for continental drift.

He soon found what he was seeking. He found that the rock formations in corresponding parts of South America and Africa were similar. Furthermore, mountain ranges on opposite continents would link up with each other if a map was drawn in which the continents had been brought together. The mountains in eastern Canada were a continuation of those in Norway and Scotland. The Sierras in Argentina linked up perfectly with the Cape Mountains in South Africa. When continents on opposite sides of the Atlantic were matched up with each other, it was as though a torn page of newspaper had been pieced back together again. The "lines of print" matched up, not just in one place, but in many different places from top to bottom.

One would think that geologists would have found Wegener's evidence convincing. Indeed, his theory of continental drift seems to have been given a reasonably sympathetic reception at first. But the scientific reaction rapidly became hostile. By the mid-1920s, Wegener's ideas began to encounter an intense antagonism. Geophysicists, in particular, subjected the theory to vehement attack. There were no known forces, they said, that could cause continents to move in the way that Wegener postulated; the entire idea was nonsense.

Wegener had suggested that perhaps tidal forces caused by the gravitational pull of the sun and the moon, and forces associated with the earth's rotation, might account for the movement of the continents. The geophysicists replied that Wegener had exaggerated their strength. If tidal forces were strong enough to move continents, they said, the same forces would halt the earth's rotation within a year. As for the rotational forces—they were too weak to have any effect whatsoever.

By the late 1920s, geologists were claiming that the supposed jigsaw-puzzle-like fit between the continents on the opposite sides of the Atlantic was imaginary. The rock formations, they said, were really not as similar as Wegener had claimed. In any case, the similarity was not sufficient to prove that the continents had formerly been connected.

Wegener's critics also attacked his credentials. He was not a professional geologist, they pointed out. On the contrary, he was an amateur who took liberties with the globe. Wegener, they said, was guilty of selecting for presentation only those facts which favored his hypothesis.

After Wegener's death in 1930, geologists and geophysicists became even more hostile to the theory. Disdain for Wegener's ideas became especially intense in the United States. If an American geologist did so much as express interest in the continental-drift theory, he was risking his reputation. Although Wegener's theory continued to have some supporters in Europe, they were in the minority.

By the 1940s, the theory was considered to have been entirely discredited. When it was mentioned at all, it was used as an example of a scientific blunder. In the view of most geologists, it was absurd to consider the idea that continents could move; all the evidence seemed to point to the conclusion that the earth was rigid.

The orthodox scientists could not, however, explain away all the evidence that Wegener had amassed. If they wanted to explain the distribution of fossils on the surface of the earth, they were forced either to resurrect the concept of land bridges or to resort to hypotheses that were even more implausible. Nor could they adequately explain why such geological features as mountain ranges should exist.

During the nineteenth century, geologists had attributed the existence of mountains to the wrinkling that had presumably taken place on the earth's surface as the planet cooled from its original molten state. But then, during the first few decades of the twentieth century, geologists began to realize that there were difficulties associated with this idea. It had become apparent that the energy released by radioactive elements adequately explained the heat in the earth's interior. There was no need to assume that the planet had once been a molten ball. Furthermore, modern theories of planetary formation indicated that the earth had been formed from cool clouds of dust. If it had changed in size at all, it could only have expanded as the radioactive energy heated it up.

In spite of its obvious deficiencies, the rigid-earth theory became geological dogma. Geologists found it easier to live with the existence of unexplained theoretical anomalies than to accept the seemingly preposterous idea that continents could drift across the face of the earth.

But then, during the 1950s, new data about the structure of the earth's crust began to appear. Before long, a revolution in geological thinking was under way.

The revolution began during the early years of the decade, when the British physicist Patrick Maynard Stuart Blackett perfected an instrument called the *magnetometer* that was capable of detecting magnetic fields only one ten-millionth as intense as that of the earth. Geologists soon realized that Blackett's device could be used to study the residual magnetism of rocks that had been formed millions of years ago.

The study of the magnetic properties of ancient rocks is a branch of geology called *paleomagnetism*. Paleomagnetic techniques can be used to determine the direction of the earth's magnetic field at the time that the rocks were formed. A magnetization that is called *fossil magnetism* or *natural remanent magnetism* is acquired by volcanic lavas when they cool and harden. The magnetization, which is aligned with the earth's field, is remarkably stable; it will remain the same if the rock is moved, or if the geomagnetic field changes. Remanent magnetism is also exhibited by sedimentary rocks. If previously magnetized particles are deposited in a sediment, they will align themselves with the earth's field. When the sediment hardens into rock, a magnetic imprint will be retained.

When geologists measured the remanent magnetism of layers

of rock in the English countryside that had been formed some 200 million years ago, they discovered that England had been situated at a latitude of 30 degrees North at that time. England is presently located at a latitude of 65 degrees. Had it migrated from one spot on the earth's surface to another over the course of 200 million years?

Although this seems to be the obvious explanation of the data, it was not the conclusion that was reached. The geologists assumed, instead, that England had remained in the same place, and that the positions of the earth's magnetic poles had shifted. At the time, it seemed easier to believe in a geomagnetic field that moved around than in shifting land masses.

Rocks of various other ages were subsequently studied, and migration paths for the North and South Poles were worked out. But the theory of *magnetic pole wandering* quickly ran into trouble. When paleomagnetic measurements were performed on rocks from other continents, the data did not quite agree with those obtained from the English rocks. For example, paleomagnetic measurements of rocks in North America produced a polar migration path that was somewhat different.

Geologists soon found that the only way to make the two paths coincide was to assume that England and North America had once been connected, and had gradually drifted apart. This suggested that there had never been any such thing as pole wandering at all. The relative motion that the rocks revealed was nothing other than continental drift.

Not all scientists were convinced. By this time, the continental-drift theory had been thought discredited for so long that many scientists were convinced that there must be some other explanation. It was suggested that paleomagnetic measurements might contain unknown sources of error. Or possibly the earth had had multiple poles at some time in the past; if it had, this could easily lead to misleading results.

The debate continued. When it began to seem that no firm conclusion could be reached, a seemingly outrageous new hypothesis was suddenly proposed. In 1962, Princeton University geologist Harry H. Hess proposed a theory of *sea-floor spreading*. According to Hess, hot material from the earth's interior was continually pushing its way to the surface through volcanic ridges on the ocean floors. As the lava cooled, it hardened into basaltic rock. But it did not remain motionless. The pressure from the material that continued to rise through the ridges pushed it outward in both directions. According to this theory, the ocean beds were moving plates. It was obvious that this would provide an explanation for continental drift. If a continent happened to be floating on top of one of these plates, it would be carried along.

But what happened to all the material that seeped through the ridges? Somehow, it had to make its way back into the earth's interior. Otherwise it would gradually build up, creating vast undersea mountain ranges, and perhaps even causing the earth's surface to expand.

Hess had an answer to that question too. Old sea floor, he said, moved back into the earth's interior at deep-sea trenches. Millions of years after it had been formed, the plates flowed back into the earth, where they melted and were absorbed.

Hess's theory sounded very plausible. Not only did it explain why there should be ocean ridges and trenches; it also provided an explanation for a fact that had been puzzling oceanographers for years. During the 1920s and 1930s, when the ocean floors had first been mapped, it had been discovered that the sediment on the ocean floors had an average thickness of only 600 meters. This indicated that the ocean floors could not be more than 100 million or 200 million years old. If they were older, the layers of accumulated sediment would be much thicker. In fact, before the ocean floors were studied, it had been expected that miles of sediment would be found.

The theory of sea-floor spreading explained why the ocean floors should be so young. They disappeared into the trenches before much sediment had a chance to accumulate. The theory seemed plausible, but it was not so easy to test. The rate of spreading was so slow that there seemed to be no hope of measuring it directly.

The following year, in 1963, the British oceanographers Frederick J. Vine and Drummond H. Matthews devised a test for Hess's theory. Vine and Matthews pointed out that there had been periodic reversals in the earth's magnetic field. These reversals, they said, should leave an imprint on the rocks of the ocean floor. The lava that emerged from the ocean ridges would be magnetized in the direction of the field. As the material spread across the sea floor, this magnetization would be retained. But if the direction of the geomagnetic field suddenly reversed, the new material emerging from the ridges would be magnetized in the opposite direction. If the reversals took place a number of times before the material disappeared back into the earth's interior, then the rocks that made up the ocean floor would exhibit a striped magnetic pattern. Bands of opposite magnetic polarity would be found between the ridges and the trenches. In addition, corresponding stripes would be found on both sides of a ridge. The theory of sea-floor spreading implied that the material moved away from a ridge in both directions. Therefore, the magnetization on opposite sides should exhibit a symmetrical pattern.

No one really knows why the magnetic field of the earth should periodically reverse itself. However, the fact that it does is well established. Paleomagnetic studies have shown that there have been 171 field reversals in the last 76 millions years. Once every 450,000 years or so—on the average—the north magnetic pole becomes a south pole, and vice versa. The reversals do not take place instantaneously. The magnetic field weakens to about 20 percent of its usual intensity, and remains at that level for 5,000 to 20,000 years before a reversal can be established.

But 5,000 and 20,000 years are brief periods on the geological time scale; they are small compared with the nearly half-million years that a field of a given orientation will persist. The finite amount of time required for the field to reverse itself does not blur the boundaries of the stripes appreciably.

Offhand, one might think that it would be practically impossible to measure the magnetization of the ocean beds. After all, it would not be easy to drill through 600 meters of sediment that might lie miles below the ocean's surface. Furthermore, if a sample could be taken to the surface, it would be hard to make sure that it had retained its original orientation.

Fortunately, nothing like this was necessary, thanks to Blackett's magnetometer. The instrument was so sensitive that investigators could measure the magnetization simply by suspending magnetometers on long cables and towing them behind ships. The magnetic orientations of the rock could then be read by scientists on the surface. Nor did the magnetization of the sediment that covered the basaltic bedrock obscure the results. Although the sediment was magnetized too, its magnetization was only about one ten-thousandth that of the basalt—not nearly enough to interfere with the readings.

During the early 1960s, the magnetization of the floors of all the earth's oceans was mapped. It was found that they were magnetized in the striped pattern that Vine and Matthews had predicted. The sea-floor-spreading hypothesis had been confirmed.

With the confirmation of Hess's theory, Wegener's concept of continental drift suddenly became respectable. By the late 1960s it was almost universally accepted. By the end of the decade, geologists took it for granted. Incorporating into it the concept of sea-floor spreading, they renamed it the theory of *plate tectonics*, and used it as a fundamental assumption when they did research.

This sounds like a story of a theory that was ahead of its time, which was finally vindicated when new evidence came to light. There is something very puzzling about the affair, however. Why should Wegener's theory have been accepted so readily in the space of a few years, after it had been reviled for decades? How could scientists dismiss it as a crackpot idea, and then accept it without a murmur a few years later?

The confirmation of the sea-floor-spreading hypothesis tipped the balance in favor of the theory of continental drift. But this does not really answer the question that I have posed. After all, most of the evidence for continental drift that was available in 1970 already existed when Wegener published his book on the subject in 1915. In 1915, scientists were aware of the apparent fit between opposing continental coastlines. They had begun to realize that the theory of land bridges was not very plausible, and they could not dispute the fact that mountain ranges matched up when continents were brought together on a map. One would think that they would at least have admitted that continental drift was a possibility.

What was it about the discovery of magnetic stripes in the ocean beds that caused them to change their minds? This hardly seems to be the most convincing piece of evidence in favor of the theory of continental drift. After all, any argument that uses the magnetic data to verify drift must proceed by a very roundabout route. One must first assume that the stripes are a record of magnetic reversals which took place millions of years ago. One must conclude, next, that this is evidence for sea-floor spreading. Finally, one must assume that evidence which implies that ocean floors move tells us that continents drift also.

The last link in the chain of arguments seems to be the weakest. After all, it would presumably be possible for ocean rock to form in one place and disappear in another without causing the continental masses to move. The evidence derived from sea-floor spreading seems to be somewhat less compelling than that which Wegener assembled during the early part of the twentieth century. If we view matters objectively, we must conclude that data derived from rock formations and studies of "pole wandering" provide more direct proof for his theory.

Nevertheless, the confirmation of sea-floor spreading seems to have represented a turning point. When this hypothesis was verified, geologists suddenly became willing to accept evidence that they had discounted for decades. Wegener was suddenly transformed from a crackpot into a scientist who was ahead of his time.

The reason that this happened was that the theory of sea-floor spreading showed how continental drift was possible. When Hess's hypothesis was confirmed, geologists realized that there existed a mechanism which could cause continental masses to move. In 1963, there was a theory that could explain the evidence that Wegener had amassed. During the 1930s and 1940s, there was not.

In all scientific fields, theory is frequently more important than experimental data. Scientists are generally reluctant to accept the existence of a phenomenon when they do not know how to explain it. On the other hand, they will often accept a theory that is especially plausible before there exist any data to support it. This is why Copernicus' heliocentric hypothesis was accepted long before the experiments that confirmed it could be performed.

Around the beginning of the seventeenth century, Francis Bacon put forth the view that scientific laws were generalizations from observed facts. A scientist supposedly performed experiments, pondered his results and then deduced a theory from what he had observed. Bacon's view of science was so influential that many people believe that this is the way science operates today.

But it isn't. Bacon's description of "scientific method" was not accurate in his own day, and it is not accurate in ours. More often than not, it is the theory that comes first. Without a theory to guide them, scientists would not know what experiments to perform. No one thought to look for magnetic stripes in the ocean floor until Hess proposed his hypothesis of sea-floor spreading. No one would have checked to see if starlight grazing the surface of the sun was bent if Einstein had not proposed his general theory of relativity. No one would have thought to perform electron-diffraction experiments if De Broglie had not put forth his theory of matter waves.

Sometimes a theory is so convincing that it is accepted even when the facts seem to contradict it. Einstein was not impressed when Kaufmann's experiments on electrons gave results that were inconsistent with the special theory of relativity. Copernicus and Kepler were not bothered by the fact that the arguments against a moving earth were difficult to answer. If a theory seems to have the ability to explain a wide variety of phenomena, if it possesses the ability to bring a large number of disparate elements together into a coherent picture of the universe, scientists will generally accept it as true. They will indeed plan to perform experiments to test it, but they will do so with the expectation that experimental data will provide a verification of the theory's validity.

On the other hand, they tend to be skeptical of observations that are not supported by theory. One of the reasons that they are so ready to discount the idea of extrasensory perception is that no one has been able to come up with a theory which can explain how ESP would operate. Admittedly, the experimental evidence in support of the existence of ESP is not very good. But that is only part of the story.

Wegener's theory of continental drift was denigrated because Wegener was unable to come up with an adequate explanation of how continents were able to move across the surface of the earth. He did amass a great deal of evidence; but without an adequate theory, that evidence did not seem very convincing.

Theories are sometimes mistaken. As a result, science sometimes blunders. Worse, the effects of the blunders occasionally persist for decades. Nevertheless, few scientists would be willing to follow Bacon's rules of "scientific method." If scientists did research the way that Bacon said they should, significant discoveries would become extremely rare. Theory occupies a pre-eminent position not because scientists tend to have their heads in the clouds, but because the scientific endeavor works best when theory is dominant.

Those scientists who do follow Bacon's prescription often fail to discover anything that is very significant. Even such scientists as Planck who try to see what experimental data imply would not get anywhere if they did not have the insight needed to see what theoretical assumptions are necessary.

On some occasions, a confrontation between theory and observation takes place. Under such circumstances, it is generally theory that wins. One such confrontation is currently happening in the field of astronomy. So far, it is theory which seems to have the upper hand.

The controversy is one that revolves around the astronomical objects known as *quasars*. The American astronomer Halton Arp has obtained observational evidence which, he claims, indicates that the accepted theoretical explanations of the quasars' brightness may be incorrect. But few astronomers accept Arp's interpretation. They tend to perceive him as a gadfly who wishes to dispute currently accepted ideas, but who is unable to come up with any good theoretical alternatives.

Quasars are bright objects that look very much like stars when they appear in astronomical photographs; the term "quasar" is short, in fact, for *quasi-stellar object*. When quasars were discovered in the early 1960s, astronomers found them very puzzling. The quasars had very large red shifts. As Hubble had emphasized in 1929, red shift and speed of recession were related. So it followed that the quasars were moving away from the earth at high velocities. This implied, in turn, that they must be very distant, billions of light-years away.

The farther away an astronomical object is, the dimmer it appears to be. This relationship holds for any luminous object; it can be applied to a candle or a light bulb as well as to a star or a galaxy. A searchlight can easily be seen at a distance of a mile, but a match flame cannot be perceived at all.

Astronomers concluded, therefore, that the quasars must be very bright. To be seen at such great distances, they had to be the most luminous objects in the universe. Some of them appeared to be shining with the brightness of a hundred galaxies. And yet the dimensions of quasars were very small. Study of astronomical data indicated that the diameter of a typical quasar was less than that of the solar system.

This discovery created problems. It was not easy to understand how something that was so small could be so bright. Some astronomers therefore proposed that perhaps the red shifts of the quasars had nothing to do with the expansion of the universe. If the quasars had been ejected from our own galaxy at high velocities, they pointed out, they could have large red shifts and yet not be very distant. The quasars, they said, looked bright only because they were nearby. This hypothetical situation would be roughly analogous to one in which a 25-watt light bulb that was very close to an observer would look brighter than a very distant searchlight.

But this theory quickly ran into trouble. Other astronomers pointed out that if quasars had been ejected from our galaxy, it was only reasonable to assume that similar objects were ejected from other galaxies also. If this was the case, it would be possible to observe quasars that were traveling toward us. These approaching quasars would exhibit blue shifts rather than red shifts. But quasar blue shifts were not observed.

Since this objection was not easily overcome, the majority of astronomers concluded that quasars were as bright and as distant as they seemed to be, and that they had existed early in the history of the universe. The last conclusion was a consequence of the fact that when astronomers look at objects that are very distant, they are also looking into the distant past. If an object is, say, 10 billion light-years away, the light that it emits must travel through space for 10 billion years before it reaches the earth.

Today astronomers generally agree that quasars are the luminous cores of young galaxies. They believe that it is possible that our own galaxy, and the galaxies near it, may have contained quasars at one stage in their evolution. A number of hypotheses have been proposed which would explain how quasars could be so bright. The most popular theory is one which proposes that quasars contain black holes which have masses a hundred million or a billion times greater than that of our sun. According to this theory, light and other kinds of radiation are given off by interstellar gas that falls into the black hole. There is every reason to believe that enormous quantities of such gas would exist in the center of a young galaxy. As this gas fell toward the event horizon of the central black hole, its gravitational energy would be transformed into heat. As it heated up, this energy would be radiated away.

This theory explains the luminosity of quasars very well. Detailed calculations show that such a mechanism could easily produce the amounts of energy that are required. The theory also explains why quasars existed only billions of years ago. Sooner or later, the supplies of gas in the galactic core would be exhausted. When that happened, the quasar would become dimmer and finally die out. The central black hole would remain, of course, but there would be no obvious sign of its presence.

If our own galaxy once contained a quasar, then it should still have a supermassive black hole in its core. If such a black hole were found, astronomers would have some evidence in support of the theory. It would not be possible to say that the theory had been proved beyond any doubt, but at least there would be some evidence to confirm it.

Unfortunately, optical telescopes cannot see into the core of our galaxy, which is hidden by clouds of interstellar dust. And of course, black holes cannot be seen directly in any case; their presence can only be inferred from the effects they have on other matter.

Astronomers have studied radio waves and infrared radiation that comes from the center of our galaxy. But there is, as yet, no conclusive evidence for the existence of a supermassive black hole. Since it is just as difficult to observe conditions in the centers of other galaxies, the black-hole theory of quasar luminosity has not been confirmed.

However, the theory does seem very plausible. It is all the more appealing because the only other theory of quasar luminosity that seems at all credible is one that is very much like it. This is the *spinar* theory, which replaces the central black hole with a spinar—a supermassive star that is spinning so rapidly that centrifugal forces prevent collapse into a black hole.

Arp expresses no opinions about the plausibility of these theories. However, he does maintain that it has not been established that quasars are as far away as they seem to be. To prove his point, he has made astronomical photographs of quasars over a period of years. He has attempted to show that quasars are often linked to galaxies that have totally different red shifts.

In some of Arp's photographs, there seem to be luminous connections between galaxy and quasar. In one case, the quasar appears to lie within the outer boundary of the galaxy itself. According to Arp, these photographs indicate that some of the current ideas about quasars should be abandoned.

His argument is a simple one. If the quasars and the galaxies are connected, then they must be located at roughly the same distance from the earth. If they are the same distance away, and they have different red shifts, then it is obviously wrong to assume that there is a relationship between quasar red shift and distance. Of course, it could be the galaxy, not the quasar, that exhibited the anomalous red shift. It is not very likely, but it is possible. However, in such a case, the validity of the red-shift/distance relationship for quasars would still be uncertain.

Arp offers no theory to explain why some quasars should have anomalous red shifts. He says that it is simply an observed fact. Most astronomers disagree. Although Arp has a few supporters, most notably Geoffrey Burbidge, director of the Kitt Peak National Observatory in Arizona, the majority refuse to believe that he has discovered any evidence that is very conclusive. They claim that the connections between the members of Arp's galaxy-quasar pairs are illusionary, and point out that perspective can make two objects seem to be very close together when they really are not.

If two astronomical objects are lined up with one in front of the other, they can seem to be associated, when the distance between them is actually very great. If a quasar is situated billions of light-years behind a galaxy, and if it lies only a fraction of a degree to one side, it can appear to be connected to the galaxy in an astronomical photograph. Arp says that statistical studies indicate that galaxies and quasars appear together too often for this interpretation to be correct. His opponents reply that Arp's statistical methods give misleading results.

Orthodox astronomers claim that the bridges of light that Arp has seen between galaxies and quasars are illusionary also. Those who have attempted to verify Arp's results say that they have often experienced difficulty in making the luminous connections visible in their own photographs. In at least one case, they add,

120 DISMANTLING THE UNIVERSE

the bridge of light turned out to be nothing more than a background galaxy that was seen edge-on.

Geoffrey Burbidge has compared the reception given Arp's ideas to the rejection of Wegener's theory of continental drift by geologists. The comparison is very apt. The evidence that Wegener presented was rejected because he could suggest no plausible mechanism that could cause continental drift to take place. Arp's evidence for the existence of anomalous red shifts has been scorned because he has no theory that would explain why such red shifts should be observed. On the other hand, Arp's opponents do have a theory; they believe that the brightness of quasars can be explained by assuming the presence of spinars or supermassive black holes.

The evidence for the existence of such objects is no more conclusive than that which Arp has presented. However, whenever there is a contest between theory and observation, it is theory which has the edge. Admittedly, theories are often overturned. But this generally happens only when a new theory appears which is more plausible than the old one. Theories are rarely discarded because anomalies of one sort or another have been discovered.

In a recent letter to *The Sciences*, a periodical published by the New York Academy of Sciences, Arp asserted that "The business of a scientist . . . seems to me to be to go out and observe real objects, see how they behave, and then induce generalizing principles from the observations." This was Francis Bacon's view of science. However, as we have seen, this is not the way that science ordinarily works. More often than not, it is theory which determines what science can observe.

It may be that Arp will eventually be vindicated, just as Wegener was. But even if he is, there will be little reason to deplore the actions of his colleagues. In rejecting evidence that cannot be fitted into any compelling theoretical scheme, the orthodox astronomers are simply behaving as scientists always have. When they say that Arp's continued search for anomalous red shifts serves no useful scientific purpose, they are undoubtedly right. Even if some of the anomalous red shifts are real, photographing more of them is not likely to lead to any increase in scientific knowledge. Only a theory that explains why anomalous red shifts would exist would tell one what to look for.

Recently, a committee of astronomers which allocates observing time on the Mount Wilson, Palomar and Las Campanas telescopes recommended that Arp be denied the use of these telescopes after 1982 unless he agreed to change the direction of his research. The members of the committee pointed out that Arp had been allocated generous amounts of observing time in the past because committee members wanted to avoid giving the appearance that they were attempting to suppress an unpopular idea. It would be possible to argue that they are suppressing Arp's ideas now. However, even if they are, their feeling that continued observations are useless when there exists no hypothesis that would allow these observations to be interpreted is probably correct. Science does not progress by observing and cataloguing strange-looking objects.

It is theory that makes observations meaningful. It imposes order upon them, and makes it possible for experimental data to be interpreted. But theories, of course, are not always correct. The insights upon which they are based are sometimes mistaken ones. After all, the intuitions of a creative scientist are no more infallible than the hunches of a compulsive gambler. As a result, acceptance of an incorrect but plausible theory sometimes causes scientists to interpret data incorrectly.

One of the most notorious examples of misinterpretation of evidence was the affair of Piltdown Man. Piltdown Man was a manufactured "fossil" that consisted of a modern human skull and the jaw of an orangutan. Both parts of the "fossil" were artificially stained to give them the appearance of great antiquity. Although the combination was a rather ludicrous one, paleontologists accepted Piltdown as a genuine human ancestor for decades. Although Piltdown Man was "discovered" in 1912, no one realized that a hoax had been perpetrated until 1953.

122 DISMANTLING THE UNIVERSE

The Piltdown "fossil" was discovered by the English solicitor and amateur geologist Charles Dawson. To this day, no one knows whether Dawson was the perpetrator of the deception, or the unwitting dupe of the hoaxer or hoaxers. There have been a number of suggestions as to who the guilty party may have been, but no one has been able to find enough evidence to convict any of the suspects.

If Dawson was not involved in the fabrication of the fossil, he cannot really be blamed for considering it an important find. After all, he was not a professional scientist. It is somewhat more surprising that some of the most eminent paleontologists of the day should have accepted the find as genuine. After all, the Piltdown "fossil" was not even a very good forgery.

When the hoax was exposed, it was discovered that the "fossil" had been constructed from 600-year-old human skull fragments and the 500-year-old jaw of an ape. The bone had been stained with paint or a paintlike substance (Britain's National Gallery suggested that it might have been ordinary Vandyke brown). The teeth in the jaw had been filed down to give them a human appearance. But this had not been done very expertly. One tooth had been filed down a little too far. When the hoaxer discovered his mistake, he had plugged the gap with a plug of plastic material that looked remarkably like chewing gum. And when the teeth were examined under a microscope, signs of artificial abrasion could easily be seen.

Once it had been proved that the "fossil" was a forgery, it became apparent that there were numerous signs of fakery that could be seen by anyone who was looking for them. For example, two of the molars in the jaw had been filed down in such a way that they were slightly out of alignment with each other. Another tooth, a canine, showed signs of having erupted only a short time before the specimen died. It had been filed down also. As a result, it exhibited signs of wear that were obviously inconsistent with its juvenile provenance.

The suspicious character of the canine had been pointed out by a dentist shortly after Piltdown was discovered. But paleontologists paid no attention to this observation. Nor were their suspicions aroused by the fact that the jaw looked like that of a modern ape. On the contrary, they unhesitatingly placed Piltdown in the human ancestral line, and made pronouncements about his evolutionary significance.

It has been suggested that the perpetrator of the hoax never expected that his little joke would be taken so seriously, and that he was afraid to reveal his deception when it was. It may never be known whether the forgery was intended as a joke or not. However, the scientists who studied the specimen have now proved astonishingly credulous.

There are undoubtedly a number of reasons why the paleontologists of the day should have accepted the find as genuine. First, although the hoax was an inexpert one in some respects, it was carried out in an elaborate manner. Genuine fossils of extinct animals were planted in the gravel bed in which the Piltdown specimen was found in order to make the skull and jaw seem to be of great antiquity. A second Piltdown specimen—a fragmentary one consisting of parts of a human skull and a molar tooth similar to the ones in the original jaw—were planted and discovered in 1915.

International rivalry and professional jealousy also played a role. At the time of the Piltdown discovery, no human fossils of any significance had been discovered in Britain. On the other hand, numerous Neanderthal specimens had turned up in France and in Belgium, and a very primitive-looking jaw had been discovered near Heidelberg in Germany. As a result, the British paleontologists had been forced to play second fiddle to their colleagues on the Continent.

When Piltdown was discovered, all this changed. British scientists stressed the fact that Piltdown, with its apelike jaw, obviously had to be much older than the "missing links" that had been discovered in France, Belgium and Germany. It appeared that the first humans had been Englishmen after all!

However, it seems apparent that these were really not the deciding factors. It is not likely that Piltdown would have been accepted as genuine if the "fossil" had not seemed to give striking confirmation to a theory of human evolution that was quite influential at the time. Shortly before Piltdown was discovered, the English anatomist Grafton Elliot Smith propounded a theory to the effect that the brain had been a kind of driving force in human evolution. According to Elliot Smith, the human brain had begun to evolve before human ancestors had acquired an erect posture, or lost their simian characteristics. Once the brain did begin to increase in size and in complexity, other modern human characteristics rapidly began to appear. When the Piltdown "fossil" was found, Elliot Smith was quite naturally delighted. He attempted to dispel doubts by insisting that the association of an apelike jaw with a very human-looking skull should not be surprising "to anyone familiar with recent research upon the evolution of man."

Today it is known that human ancestors began to walk upright while their brains were no larger than that of a chimpanzee. But in 1912, the "brain first" theory seemed a very attractive one. After all, the complexity of the human brain seems to be what most distinguishes us from other mammals. With no evidence to the contrary, what would be more natural than to believe that the brain somehow led the way in evolution?

When Piltdown came to light, some scientists were skeptical at first. They suggested that the skull and jaw were not fossils of the same individual. They must have come together accidentally, they said. But then Elliot Smith threw his considerable scientific weight behind the paleontologists who had pronounced Piltdown to be genuine. Before long, other important members of the British scientific establishment had followed. By the time it was discovered that they had been wrong, the experts who had insisted that the specimen be accepted as genuine had all died, or retired.

The Piltdown discovery sent science blundering off in the wrong direction. Since a hoax was involved, it could be argued that this was not a typical case. It is obvious, however, that scientists are perfectly capable of making monumental mistakes even when there is no trickery involved. As one might expect, it is normally theory that causes the blunders to be made. A good example is provided by the discovery, in 1859, of an imaginary planet.

In 1859, it had long been known that the planets in the solar system did not travel along orbits that were perfectly elliptical. By this time, it had been realized that Kepler's law was an approximation, that the gravitational attraction of one planet for another created perturbations that caused the orbits to become slightly irregular. This effect is particularly noticeable in the case of the planet Mercury. Mercury's orbit swivels around the sun so that the point of closest approach, the *perihelion*, is not always in the same place.

Mercury's perihelion swings around the sun at the rate of $1°33'20''$ (1 degree, 33 minutes and 20 seconds) per century. Of this, $1°32'37''$ can be explained by Newton's law of gravitation; calculations show that the other planets should perturb Mercury's orbit by this amount. The remaining 43 seconds of arc are due to relativistic effects; this discrepancy is explained by Einstein's general theory of relativity.

In 1859, the French astronomer Urbain Jean Joseph Leverrier used Newton's theory to calculate the effects that the other planets would have on Mercury's orbit. His calculations indicated that there existed a discrepancy of 38 seconds of arc, which is very close to the modern value of 43 seconds. The error was only a little more than 10 percent.

There was no way that Leverrier could have dreamed that Mercury's deviation from its expected orbit would eventually be explained by Einstein. As a result, he made what seemed to be a very natural assumption. He assumed that the deviation was caused by an as-yet-undiscovered planet. Leverrier called this planet "Vulcan," and hypothesized that it lay inside the orbit of Mercury. He calculated that a planet halfway between Mercury and the sun would have just the right effect if it had two-thirds of Mercury's mass.

Leverrier had every reason to believe that his methods would lead to a correct result. In 1846, he had performed similar calculations on the orbit of the planet Uranus. His predictions had led to the discovery of Neptune in the very same year.

When Leverrier announced that he had deduced the existence of an inter-Mercurial planet, a French physician and amateur astronomer named Lescarbault announced that he had sighted the planet some months earlier. When viewed in his telescope, it had seemed to be a dark spot moving across the disk of the sun.

During the next few years, Dr. Lescarbault's discovery was confirmed by famous astronomers of a number of different nationalities. They apparently had little difficulty seeing a planet which did not exist. Furthermore, searches of astronomical records revealed that other observers had sighted the planet as early as 1802. Soon the existence of Vulcan was considered to be so well confirmed that textbooks on astronomy began to assign it a place in the solar system that was as secure as that of Mercury itself.

But astronomers soon found that it was not easy to work out the details of Vulcan's mass and orbit. Sometimes it didn't even seem to be there when they looked for it. At first, this didn't trouble them much. No one had thought that observations of Vulcan would be easy. It was to be expected that a planet so close to the sun would often be blotted out by the sun's light. After all, Mercury itself was difficult to observe at times.

As additional searches were made for Vulcan, the situation rapidly became even more confused. Astronomers looked for Vulcan during solar eclipses in the hope that the blotting out of the sun's light would allow them to observe it clearly. Sometimes they saw (or thought they saw) it, and sometimes they didn't. And when Vulcan was observed by different astronomers, the results did not always agree. In 1878, two noted astronomers reported that there existed not one, but two planets inside Mercury's orbit. However, the two planets reported by one astronomer did not seem to follow the same paths as the two reported by the other. Meanwhile, other astronomers who had been observing at the same time saw nothing at all.

The problem of the orbit of Vulcan attracted even greater interest when the perfection of dry-plate photography provided astronomers with a tool that was easier to use than those which had previously been available. After the problems associated with photographing faint bodies near the sun were solved in 1900, photographic searches for Vulcan were systematically conducted whenever an eclipse of the sun occurred. But observations during the eclipses of 1900, 1901 and 1908 revealed nothing. Although reports of visual sightings did not cease entirely, most astronomers came to the conclusion that Vulcan did not exist.

But the question was not completely settled until 1916, when Einstein published his general theory of relativity, and explained the discrepancy of 43 minutes of arc. Within a few years, speculation about the possible existence of Vulcan, which had lingered on up to this time, ceased entirely. Theory had explained the irregularities in Mercury's orbit, and had made the planet Vulcan unnecessary. The theory of 1859 had suggested that Vulcan might exist, and it was the theory of 1916 which proclaimed that the case was closed.

CRACKPOT THEORIES

WE TEND TO HAVE a great deal of confidence in scientific knowledge nowadays, and there is every indication that this faith is justified. Not only has twentieth-century science led to astonishing technological advances; it has also allowed us to gain an understanding of such things as the nature of the genetic code, the behavior of the fundamental constituents of matter and the evolution of the universe. In our time, the acquisition of scientific knowledge has proceeded at an ever-increasing rate. Government and private support of science has grown to such a degree that the majority of the scientists who ever lived are alive today.

But if theories can be based on mistaken insights, and if this can lead, in turn, to the misinterpretation of data, then one would think that scientific knowledge would be full of errors. In such a case, it would not be possible to have confidence in anything, and the only reasonable attitude toward scientific endeavor would be one of skepticism.

It is obvious that such skepticism would not be justified. Modern science is too successful. Its success, therefore, is a paradox that needs to be explained. Blunders such as the acceptance of the genuineness of the Piltdown fossil and the discovery of Vulcan are inevitably corrected sooner or later. But what are the self-correcting mechanisms in science? Why should they be so accurate?

Or perhaps we should phrase the question a little differently, and ask why it is that a scientist cannot make up any kind of theory that he desires. Although the invention of a theory is an act of creative imagination, some theories are right and some are wrong. There are others that are too ridiculous even to be considered. What, exactly, is the difference?

A good place to begin might be in 1903. That was the year in which one of the classic blunders of twentieth-century physics took place. In 1903, the distinguished French physicist René Blondlot created a sensation by announcing his discovery of a new type of radiation, N rays.

Blondlot, who was a member of the French Academy of Sciences, was anything but a crackpot. He was a respected scientist and a meticulous experimenter. Nevertheless, he spent years investigating a type of radiation that did not exist. Nor was he the only one to be deceived. After he made his discovery, numerous other scientists confirmed his results by reporting that they had detected N rays as well.

After X rays were discovered in 1895, Blondlot became involved in attempts to answer a question that had been puzzling scientists around the end of the nineteenth century: were X rays made up of particles, or were they a form of electromagnetic radiation? During the course of his experiments, Blondlot discovered that something was causing an electrical spark that was jumping between two wires to become brighter. At first, he thought that the X rays were causing this effect. But further experiments demonstrated that this was impossible. For example, whatever it was that was producing the effect seemed to be deviated by a quartz prism that was placed near the spark gap. Previous experiments had shown that X rays were not bent when they passed through the quartz.

At this point, Blondlot made an intuitive leap. It was not of the magnitude of those made by Einstein, or by Kepler or Galileo, but it was a conceptual leap nonetheless. Blondlot realized that if

X rays were not brightening the spark, then some other kind of radiation was. Furthermore, it had to be a kind of radiation that had not previously been discovered.

Blondlot continued his experimentation. He replaced visual examinations of the spark with exposures on photographic plates. He conducted other experiments which seemed to indicate that N rays could cause phosphorescent strips to glow brighter. He found that N rays could be focused with an aluminum lens, and that there were apparently a number of different N-ray wavelengths.

Shortly after Blondlot published his results, other scientists began to enter the new field of research. Physicists, physiologists and even psychologists joined in the investigations. It was found that wood, paper and thin sheets of such metals as tin, silver and gold were transparent to the new type of radiation. On the other hand, such substances as water and rock salt were opaque to it. N rays passed through iron sheets with ease if the material was not too thick. On the other hand, they were effectively blocked by pieces of wet cardboard.

Soon it was discovered that N rays were emitted by the human body, especially from the nerves and muscles. A member of the French Academy reported that he had detected N-ray emissions from corpses, and it was suggested that the radiation could be used to examine the human body for medical purposes. It was pointed out, for example, that there was no reason why the radiation could not be used to obtain an outline of the heart.

But some experimenters found themselves unable to reproduce the reported experimental results. When they tried to produce and detect N rays in their own laboratories, they encountered nothing but failure. Soon, some of them began to entertain doubts about the reality of this new type of radiation.

One of these skeptical scientists was the American physicist Robert Williams Wood. In 1905, Wood visited Blondlot's laboratory to observe the experiments for himself and found that he could not see the fluctuations in the brightness of the spark that Blondlot claimed to be observing. Wood suspected also that the photographic experiments had not been set up in such a way as to eliminate all sources of error.

Naturally, Wood could not be sure about his suspicions. In the end, he resorted to a piece of trickery in order to perform a convincing test. When Blondlot darkened the laboratory in order to perform an experiment with an aluminum prism and a phosphorescent strip, Wood surreptitiously removed the prism from its mounting. Blondlot went ahead with the experiment and reported that the nonexistent prism was causing the N rays to be deviated.

Even after Wood demonstrated that N rays were an illusion, many French scientists continued to support Blondlot. Some proponents of the N-ray theory went so far as to claim that only French scientists had the visual sensitivity to see the evidence for the radiation. The perception of Anglo-Saxon scientists, they said, had been dulled by exposure to English fog, while that of their German colleagues had been blunted by excessive consumption of beer.

But eventually even the French began to suspect that the evidence for the existence of N rays was spurious. Interest in the phenomenon subsided in 1906 when Blondlot refused to participate in a test for the existence of the radiation that had been proposed by the French journal *Revue scientifique.* "Let each one form his personal opinion about N rays, either from his own experiments or from those of others in whom he has confidence," Blondlot said.

It must have sounded like a curiously unscientific statement. Science, after all, is not supposed to be a matter of personal opinion; it is supposed to concern itself with results that can be duplicated anywhere. Science exists to explain phenomena that anyone with the proper equipment and sufficient expertise can detect. The demand that this be possible is one of the factors that separate plausible hypotheses from erroneous theories.

In a sense, scientific theories are free creations of the human imagination. However, there are constraints upon the kinds of theories that can reasonably be proposed. Science, after all, exists to explain natural phenomena. When it attempts to deal with phenomena that do not exist, the blunder that started it all can generally be exposed in a relatively short period of time.

This is exactly what happened in the case of N rays. It is true that it took somewhat longer for astronomers to discover that the planet Vulcan was a chimera. But at the time, there were reasonably good theoretical grounds for thinking that an inter-Mercurial planet might exist. Even in that case, doubt began to arise when it proved that repeatable observations were simply not possible.

This, by the way, is the reason that scientists tend to be skeptical about stories concerning UFOs, and about mysterious events which are supposed to take place in that area of ocean called the Bermuda Triangle. Since there is no solid evidence to indicate that UFOs are real, the study of such "objects" is considered to be pseudoscience. No one has ever discovered any artifact of extraterrestrial origin, and the majority of UFO sightings can be attributed to natural causes. Most likely, all of them could be explained away if scientists only possessed sufficient information. It is true that some UFOs have shown up on radar. But these sightings took place when radar technology was not as advanced as it is now, and could have been caused by such phenomena as meteors.

Similarly, there is not a shred of evidence which would indicate that anything out of the ordinary happens in the Bermuda Triangle area. It is true that ships and airplanes have disappeared in that region. However, airplanes do sometimes crash and ships do sometimes sink. Furthermore, careful research has revealed that some of the ships that are supposed to have disappeared in the area never existed in the first place. It seems that some tales of mysterious events have been repeated so many times that they have taken on an aura of reality, even though they lacked any foundation whatsoever.

Scientists are sometimes stereotyped as individuals who are overly skeptical. It is thought that they tend to be too "closed-minded" to believe anything that is not in accordance with their theories. In reality, there is some truth in this accusation. However, their skepticism is not a vice but rather a virtue. If they did not possess this quality, it is likely that we would all wind up believing in UFOs, astrology, the Bermuda Triangle and such phenomena as N rays and inter-Mercurial planets as well.

Scientists blunder in a variety of different ways. Sometimes they observe phenomena that are not there. Sometimes they observe something that is real, but interpret the evidence incorrectly. A good example of the latter is provided by an episode that took place during the 1960s, when scientists all over the world subjected a novel substance called "polywater" to intensive study.

In 1962, a Soviet scientist, Nikolai Fedyakin, discovered a method for isolating a substance which appeared to be water, but which seemed to have some very surprising properties. Research on the subject was quickly transferred to the Institute of Surface Chemistry in Moscow, where it was confirmed that the "modified water" did indeed exist.

Interest in the subject spread rapidly. By the mid-1960s, hundreds of scientists throughout the world were investigating the substance, which was now universally known as "polywater." The researchers found that polywater had strange properties indeed. It was about one and a half times as heavy as ordinary water, and it did not boil or freeze at the same temperatures. Furthermore, it was considerably more viscous. That is, it flowed less readily; it seemed in this respect to be less like water than like a substance such as petroleum jelly.

Hundreds of research papers were published on both sides of the Atlantic. Theories were proposed to explain polywater's curious behavior, and scientists began to speculate about its possible uses. The Soviet scientists who had initiated the research reported that they had determined that polywater had a chemical formula of H_8O_4 (rather than the usual H_2O), while an American scientist warned that stringent precautions should be observed by those who worked with the substance. There was a possibility, he said, that polywater samples could act as "seeds" that would convert the water in the oceans into polywater too. If this happened, he claimed, the earth might be transformed into a barren planet like Venus.

The U.S. Department of Defense became interested in the possible military uses of polywater, and it sponsored research on the subject. Meanwhile, sensational stories began to appear in the news media. "A few years from now living room furniture may be made out of polywater," said an article in *The Wall Street Journal*. "Scientists Growing Wet, Creepy Water," proclaimed a headline in *The Washington Post*. There was speculation that a Nobel Prize would be awarded for the discovery of the substance, and there were the inevitable arguments over priority. A British scientist claimed that polywater had actually been discovered by his father in 1928.

By 1970 it had become apparent that polywater was no more real than N rays had been. It was nothing more than ordinary water which contained a high level of impurities. The impurities had been difficult to detect because the methods that had been developed for isolating polywater were of such a nature that only very small quantities could be produced. This had made analysis difficult and had caused scientists around the world to participate in a collective delusion.

However, there was no single, crucial experiment which demonstrated that polywater was not real. In fact, when skeptics first pointed out that polywater samples contained high levels of impurities, they were not believed. "The polywater *you* prepared may be impure, but that doesn't happen in *my* laboratory" was a typical response of polywater researchers.

The polywater craze finally ended, not because the existence of the substance was disproved, but because the theoretical structure that had been built up had become so fantastic. The more polywater was researched, the more remarkable qualities it seemed to have. In the end, polywater seemed to exhibit so many bizarre properties that scientists began to wonder if perhaps the idea that it was nothing but a solution of impurities might not be right after all. Once the members of the scientific community began to entertain doubts, it didn't take long before the theoretical enterprise that had been built up around the substance fell crashing to the ground.

It is easy to make a mistake and to observe something that does not really exist. It is more difficult to explain a chimera theoretically. Sooner or later, inconsistencies will be noticed. It will be difficult to work out a plausible theory. If one is found, experimental results are not likely to confirm it. When scientists investigate a phenomenon that is real, theoretical explanations gradually become more coherent. But when they research an illusion, the theories that are produced are likely to take on an *Alice in Wonderland* character.

There is a real world, after all. When scientists describe it, they may argue with one another at first. In most cases, however, a consensus gradually develops. Achieving a consensus about an imaginary substance like polywater is more difficult, for there is nothing real to agree upon. Something that is a product of scientists' imaginations is not likely to appear the same to one individual as it does to another. This inevitably leads to chaos. Science does have its subjective elements. A theory can triumph because it is based on especially imaginative insights. A theoretical scientist is often influential because he possesses a theoretical vision that his colleagues find persuasive. But however imaginative he is, he is dealing with objective reality. Theories may be invented rather than deduced. However, scientific theories are invented to explain the things that happen in the world around us.

When a theory describes an imaginary world, it is generally not taken very seriously at all. A theoretician cannot invent anything he chooses. If he attempts to do so, he is generally dismissed as a crackpot. Not surprisingly, this is exactly the reception that has been given to Immanuel Velikovsky's book *Worlds in Collision*.

When *Worlds in Collision* was published in 1950, scientific reaction against it was so intense that the publisher, the Macmillan Company, was forced to transfer the book to Doubleday in order to avoid a boycott of Macmillan's textbooks by scientists who taught in American universities. These scientists were outraged by the fact that a reputable scientific publisher would bring out such a book, and were dismayed to find that the theory expounded in the work was being taken seriously.

During the late 1950s and early 1960s, astronomers made a discovery that was touted, in some quarters, as a vindication of Velikovsky's ideas. They discovered that the surface temperature of the planet Venus was much higher than they had previously thought. In fact, it was even higher than they had previously considered possible. The surface temperature of Venus turned out to be approximately 460 degrees Celsius (about 860 degrees Fahrenheit), a temperature considerably higher than the melting points of such common metals as lead, tin and zinc. This was so surprising that when studies of radio emissions from Venus which began in 1956 indicated that temperatures above 400°C existed on Venus, astronomers refused to believe the results. It was only after Soviet *Venera* space vehicles entered the Venusian atmosphere during the 1960s and radioed back temperature and pressure readings that the existence of such astonishingly high temperatures was accepted.

Now, it so happened that high temperatures were predicted by Velikovsky's theory. According to Velikovsky, Venus, which had originally been ejected from the planet Jupiter, had passed close to the sun at one point in its history. The sun had heated Venus to a state of candescence. Since there had not been enough time for Venus to radiate away all its heat, Velikovsky had concluded that it must still be very hot.

Did this observational confirmation of the theory cause astronomers to reverse themselves and accept the validity of Velikovsky's ideas? Of course not. Velikovsky's theory was considered to be crackpot when it was published in 1950, and it is still considered to be crackpot today. The "verification" of the theory by measurements of the surface temperature of Venus has not altered things one whit. The fact that Velikovsky made a correct prediction in this case is passed off as nothing but coincidence.

As a matter of fact, the verification is really not all that convincing. Velikovsky's theory says that Venus should be hot, but it does not say exactly how hot the planet should be. Unlike most orthodox scientific theories, it fails to give a quantitative prediction. Furthermore, some of its other predictions are not confirmed. Velikovsky's theory implies that Venus must be cooling off. But observations made over a period of years indicate that it is not. Furthermore, observations of the clouds of Venus indicate that they have precisely the temperature that would be expected due to heating by the sun.

According to Velikovsky, the clouds of Venus are made of hydrocarbons (chemical compounds of hydrogen and carbon). However, it was established in 1973 that these clouds are made up primarily of sulfuric acid vapor. Nor have any hydrocarbons been found in the Venusian atmosphere, which is 93 percent carbon dioxide and 7 percent nitrogen, water vapor and carbon monoxide. Traces of other gases are also present, but hydrocarbons are not.

Some of Velikovsky's ideas about the composition of the Venusian atmosphere sound too ludicrous to be credible. He claims that the manna that fed the Children of Israel in the wilderness was composed of carbohydrates that fell from Venus when the planet passed close to the earth. It is not clear whether he believed that hydrocarbons and carbohydrates were both present in the Venusian atmosphere, or whether he was laboring under the impression that they were the same thing. Velikovsky also suggests that an earlier passage of Venus near the earth was responsible for the Biblical plagues that persuaded the Egyptian Pharaoh to let the Israelites go in the first place. In particular, he suggests that some of the vermin that plagued Egypt—the flies in particular—fell to the earth from Venus.

It sounds rather implausible. But the Copernican hypothesis was also thought to be implausible when it was first proposed. So were some of Einstein's ideas. What right do we have to label Velikovsky's theory as "crackpot"? For that matter, how can we be sure that the theory is not correct in its broad outlines, even though it fails to give correct predictions about the atmosphere and clouds of Venus? Cannot we forgive Velikovsky for his silliness about the Venusian flies, and take some of his other ideas seriously?

I think that the answer must be no. Velikovsky was a crackpot. There is no other accurate way to describe him. His theories were so absurd that no reputable scientist would be able to take them seriously. There is a world of difference between a crackpot theory and one that is scientific. The ideas expressed in *Worlds in Collision* clearly belong in the former category.

The difference between crackpot and scientific theories has little to do with correctness. Theories can be wrong and still be scientific. In fact, one of the ways that science progresses is by proving reasonable-sounding hypotheses to be incorrect. The more science knows about what is not true, the better its understanding of what might be true.

"Crackpot" is not an epithet that is applied to any theory which a scientist happens not to like. When Einstein published his theories of relativity, there were many eminent scientists who refused to believe that his ideas could be correct. But they did not dismiss Einstein's ideas as "crackpot." They took Einstein's theories seriously, and attempted to meet them with reasoned scientific objections. In the end, Einstein proved to be right. But even if it had turned out that he was mistaken, scientists would not have classed him with the flat-earth theorists, or with those who believe that manna fell from the planet Venus.

The difference between scientific and crackpot theories does not have much to do with experimental confirmation either. There have been many scientific theories which could not be subjected to experimental tests until years or decades had passed. Confirmation of Newton's law of gravitation, for example, was not obtained until long after Newton was dead. On the other hand, crackpot theories sometimes produce predictions that accidentally turn out to be correct. Although Velikovsky made no quantitative predictions about the temperature of Venus, he did maintain that the planet was hot at a time when most astronomers were certain that it was relatively cool.

It would be foolish to claim that scientists have always been successful in distinguishing between scientific and crackpot ideas. It is obvious that they have not. Meteorites were once thought to be a superstition; scientists laughed at the idea that stones could fall from the sky. Wegener was laughed at when he proposed that continents could drift across the face of the earth. On the other hand, the concept of polywater was taken quite seriously for a number of years. Nevertheless, it is possible to identify most of the cranks most of the time. Crackpot theories generally share certain characteristics. In order to see what these characteristics are, it might be well to look at Velikovsky's theory in greater detail.

Velikovsky seems to have believed that ancient myths were to be taken literally and, furthermore, that mythical events had natural explanations. His theory of the origin of Venus is derived from a Greek myth about the birth of the goddess Athene, and his ideas about the behavior of the planet thereafter are derived from the myths of a number of different peoples, including those which were written down in the Old Testament. According to the ancient Greeks, Athene was born from the brow of Zeus. It seems that at one time, Zeus had lusted after a female Titan named Metis. Although Metis changed her shape numerous times in order to elude him, Zeus finally caught and raped her. According to Greek myth, the unions of the gods were always fertile. Not surprisingly, Metis became pregnant. Shortly thereafter, an oracle foretold that if Metis conceived, she would bear a son who would be fated to depose Zeus. In order to avert this fate, Zeus caught Metis again and swallowed her.

This was the end of the Titaness, but not of the child. When the term of gestation was completed, Zeus was seized by a raging headache. The pain was so intense that the god's cries echoed throughout the firmament. Hearing the cries of agony, the god Hermes rushed up and split open Zeus' skull with a wedge. As soon as the breach was made, Athene sprang forth from Zeus' brow, wearing full armor.

Now, Zeus is identified with the planet Jupiter. In fact, "Jupiter" is the Roman name for Zeus. Velikovsky concluded, therefore, that the myth told of an astronomical event. Specifically, it implied that an astronomical body had been ejected from Jupiter's interior. According to Velikovsky, this body was nothing other than the planet Venus.

It so happens that Aphrodite, not Athene, is the Greek goddess who is ordinarily identified with Venus. However, Velikovsky claimed that there were reasons for believing that this identification was incorrect. Athene represented the planet Venus, he claimed, while Aphrodite was really a symbol of the moon.

In Velikovsky's theory, then, Venus was originally a comet that was expelled from Jupiter around the middle of the second millennium B.C. During its life as a comet, Velikovsky said, Venus passed quite near the earth on several occasions. Eventually it collided with Mars. It was this latter collision which forced it into the stable, nearly circular orbit that it now occupies.

The passages of Venus near the earth were presumably responsible for the catastrophic events that are described in the Old Testament and in the mythologies of various ancient peoples. Not only did Venus cause the plagues that Jehovah visited upon the Egyptians and produce the manna that fell in the wilderness, it also caused the parting of the Red Sea. A later collision between Venus and the earth stopped the earth's rotation entirely and caused the sun to stand still for the Israelite leader Joshua. After Joshua had won his battle, the forces exerted by Venus caused the earth to start spinning again.

The most striking thing about this theory is that many of the facts known to orthodox science are ignored. Velikovsky does not seem to be bothered by the fact that no astronomical body—large or small—has ever been observed to be expelled from Jupiter. Nor does he make any suggestions as to where the enormous energy required to eject a mass the size of a planet might come from (the required energy would be roughly equivalent to that radiated by the sun in a year), or explain why Venus would not burst into fragments when this violent process took place.

Velikovsky says that Venus collided with the earth on at least two different occasions around the middle of the second millennium B.C. If the first collision is to explain the events described in the Old Testament, then Venus must have remained in the vicinity of the earth for a period of months. Obviously, a certain amount of time had to pass between the first of the plagues and the falling of the manna in the wilderness. But it is not easy to imagine what the forces could be that would keep two planets together for this period of time, and then allow them to move apart. When an astronomical body orbits around another, the arrangement tends to be permanent.

Velikovsky assumes that there were grazing collisions between the earth and Venus on two separate occasions. He says that there was also a collision between Venus and Mars, and another between Mars and the earth. Any one of these events would be extremely improbable even if planets could stray from one orbit to another the way that Velikovsky says they did. It is inconceivable that collisions of this kind could take place on three or four separate occasions. The volume of space in the solar system is so large and the diameters of the planets so small by comparison that bodies whose orbits do intersect collide only rarely.

For example, it is estimated that there are 750 to 1,000 asteroids which have orbits that intersect that of the earth. Yet a collision with one of these bodies takes place only about once every 250,000 years. Venus is larger than a typical asteroid. So if Venus did move into a trajectory which passed that of the earth, a collision might be a little more likely. But the probability would still not be very large. According to a calculation performed by the American planetologist Carl Sagan, the chance that such a collision would take place in any given year is about one in 30 million.

Improbable events sometimes happen. If an object the size of Venus were ejected from Jupiter, there would be some finite chance that it would collide with the earth. However, assuming that such an improbable event was followed by a number of other, equally improbable events strains the bounds of credibility.

If planets repeatedly collide with others, how do they survive the encounters? If we assume that they don't actually crash, but simply pass within a thousand or so kilometers of each other, this really doesn't solve the problem. If two bodies the size of the earth and Venus moved this close to each other, tidal forces would cause both to disintegrate.

Velikovsky claims that colliding planets do not destroy each other because they are cushioned by magnetic fields. This is nonsense. Venus may not have any magnetic field at all. Measurements made by the *Venera 4* space vehicle in 1967 have established that if Venus does possess a magnetic field, it can have no more than one ten-thousandth the strength of the earth's. The earth does have a field, but it is not very strong compared with the fields that can be produced in the laboratory. The earth's field is too weak even to move such small objects as nails or safety pins. It can move a compass needle, but only because the needle is magnetized itself, and delicately balanced. It is not conceivable that the earth's feeble field could cushion the impact of a planet.

Velikovsky's theory sounds quite fantastic. But then, Einstein's ideas sounded fantastic too when they were first proposed. What, exactly, is the difference between Velikovsky's theory and Einstein's? How can we accept one set of fantastic ideas while refusing even to consider the other?

The difference is that Velikovsky ignored well-established scientific laws, while Einstein did not. Velikovsky did not attempt to show that accepted ideas about the behavior of astronomical bodies were mistaken; he acted as though all the accumulated knowledge of the astronomers did not exist.

In particular, Velikovsky did not attempt to explain how a body the size of Venus could be expelled from Jupiter, or why statistically improbable collisions should happen over and over again. In working out his theory, he ignored well-established principles in physics, celestial mechanics and the mathematical theory of probability. Velikovsky's theory depends upon the existence of unknown sources of energy, on tidal forces that fail to act, on mysterious forces that draw planets together and on magnetic fields that become miraculously strong.

The special theory of relativity, on the other hand, doesn't confine itself to predictions concerning the behavior of objects that are moving at high velocities; it also explains why Newtonian physics should work so well under ordinary circumstances. Einstein did not ignore the laws of Newtonian physics; on the contrary, he showed that they should remain valid at low velocities.

When objects move at velocities that are a considerable fraction of the speed of light, relativistic effects can be quite large. But when they move at ordinary terrestrial velocities, these effects are generally too small to be measurable. For example, if an automobile is moving at a velocity of 30 kilometers per hour, the contraction in its length will be approximately equal to the diameter of an atom.

The physicists who developed quantum theory did not ignore previous knowledge either. When they realized that light sometimes behaved as if it were made up of particles, they did not attempt to throw the old wave theory overboard. They realized that each of the conceptions of the nature of light was valid under certain circumstances. They understood that the old and the new descriptions had somehow to be reconciled.

Scientific theories are often overthrown. But when this happens, they are not simply tossed on the scrap heap. The new theory must be able to account for the successes of the old one. It must explain why the theory to be discarded managed to account for observed phenomena as well as it did.

Sometimes the old theory is retained, and used in circumstances where it is adequate. For example, no astronomer in his right mind would use the equations of general relativity to calculate the orbit of a comet. The Newtonian law of gravitation is much simpler, and the relativistic corrections are so small that they can be disregarded in this case. Similarly, the wave theory of light is still very useful. It does not explain the emission of light from atoms, which is a quantum process, but it is perfectly adequate for describing the propagation of light under most circumstances.

Velikovsky's supporters sometimes charge that orthodox scientists dismiss the ideas propounded in *World in Collision* because they will not admit that Velikovsky successfully challenged orthodox theories. Nothing could be further from the truth. Velikovsky did not challenge anything. On the contrary, he behaved as though modern science did not exist. He based his speculations on myth alone, disregarding numerous well-known facts. Velikovsky has to be considered a crackpot because he operated outside the bounds of science.

If the term "crackpot" is used in this sense, we should probably consider the N-ray and polywater theories to have initially been mistaken, but not crackpot, ideas. After all, attempts were made to interpret the illusionary N-ray and polywater phenomena in terms of accepted scientific ideas. Skepticism among scientists became general only when these theories became so convoluted that they began to take on crackpot proportions. When one attempts to explain something that isn't there, it is inevitable that one will pass beyond the bounds of credibility sooner or later.

According to this definition, the idea of a planet inside the orbit of Mercury was not crackpot at all. On the contrary, it was a consequence of the scientific theory that existed in 1859. Astronomers had every right to expect that such a body would eventually be found. It was only when a better theory was discovered that they knew that they had been mistaken.

A theory can be incorrect and still be part of ordinary science. The weeding out of erroneous ideas is part of the scientific endeavor. But scientists do not ordinarily devote much energy to attempts to prove that crackpot theories are incorrect, for the simple reason that these theories cannot be fitted into the scientific enterprise in the first place.

There are some theories which are so plausible that they gain acceptance long before experimental confirmation is available. Crackpot theories represent the other side of the coin; they are so implausible that they are generally dismissed without a thought even when observational data seem to offer some confirmation.

If a crackpot theory is one that is untrue and patently ridiculous, how would we classify a theory that is almost certainly false, but that doesn't ignore the things that are known to science?

Wouldn't we have to consider such a theory to be "scientific"? Of course we would. And as a matter of fact, such theories exist. Furthermore, scientists often discuss them quite seriously, even when they doubt that they can possibly be correct.

According to one of these theories, known as "the many-worlds interpretation of quantum mechanics," there exist an infinite number of alternative universes which are much like the "parallel worlds" of science fiction. If the many-worlds interpretation is correct, the universe is splitting into numerous replicas of itself at every instant. Furthermore, every time this happens, everything that the universe contains is duplicated also.

According to this theory, there have been an infinite number of such splittings in the past. Consequently, I have an infinite number of counterparts in alternative universes. Furthermore, every time the universe bifurcates in the future, more duplicate mes will be created. My counterparts do not always live lives that resemble mine. In fact, many of them are already dead. And since anything that can happen has happened to at least one of them, I can be sure that some of these other Richard Morrises are major-league baseball players, while others head large corporations, or are serving terms in prison.

If the theory is true, there must be countless other universes in which I never existed. Furthermore, there are universes in which the Roman Empire never fell, and universes in which the Confederacy won the American Civil War. And of course there are universes in which the earth was destroyed by some great astronomical catastrophe before *Homo sapiens* evolved, and universes in which the earth is populated by intelligent dinosaurs. Finally, there are an infinite number of universes in which life failed to evolve, or in which stars and planets were not able to form.

The many-worlds interpretation of quantum mechanics was proposed by the American physicist Hugh Everett III in his Princeton University doctoral dissertation in 1957. According to Everett, it is not necessary to accept the idea that quantum mechanics is nondeterministic. It is possible to assume, instead, that the universe bifurcates whenever a quantum event takes place.

An example should make this somewhat clearer. Radium 224 is one of the hundreds of known radioactive isotopes. An atom of radium 224 decays by emitting an alpha particle. When it does, it is transformed into an atom of radon gas.

It is impossible to predict just when any individual atom of radium 224 will decay. Since the emission of an alpha particle is a process that is governed by quantum-mechanical laws, it is possible to speak only of probabilities. The best we can do is to say that after 3.64 days, there will be a 50-percent chance that any given atom has decayed. This is the same thing as saying that half of the atoms in any given sample will have decayed at the end of this time. This is the origin of the term *half-life*.

Or at least this is the standard interpretation of radioactive decay. In the many-worlds interpretation, this is not what happens at all. According to Everett's theory, the atom of radium decays at *every* moment of time, and each one of these events corresponds to a different universe. There is one universe in which the atom decays after 0.5 second, another in which it decays after 1.81 seconds, another in which it decays after five weeks and so on. Since there are an infinite number of different possibilities, a single atom of radium, by "deciding" to decay or not to decay at any given instant, can create an infinite number of new universes.

Radium 224 is only one of many different varieties of radioactive atom, and radioactivity is only one of numerous processes that are described by the laws of quantum mechanics. If Everett's theory is correct, every particle in the universe causes the universe to split in two every time it encounters a situation in which it has a "choice." When an atom emits a photon of light, the universe bifurcates. When that photon encounters another particle, the universe bifurcates again. Or perhaps "bifurcate" is inadequate to describe what happens. The photon does not cause the universe to split once; it causes that splitting to take place over and over again.

Everett's theory is so outrageous that even someone accustomed to encountering fantastic ideas in science is likely to find it hard to believe. Anyone who accepts the theory is forced to conclude that he is split into uncountable replicas of himself whenever a quantum event takes place in galaxies that are billions of light-years away.

The implications of the many-worlds interpretation sound so absurd that one would think it would be easy to disprove the theory. Surprisingly, this turns out not to be the case. There does not exist the slightest bit of experimental evidence that can be used against Everett's theory.

There may not be any way that such evidence could be obtained. The many-worlds interpretation is just that—an interpretation. It is one of the possible alternatives to the interpretation of quantum mechanics that was developed by Bohr and his colleagues at Copenhagen. It cannot be disproved because it makes use of the standard mathematical formalism that is used by the more orthodox version of the theory. In a sense, every experiment that confirms orthodox quantum mechanics confirms Everett's many-worlds interpretation too. After all, one cannot see the bifurcations that presumably take place. We can only perform calculations for the universe that we inhabit. If the other universes are there, their existence does not affect the half-life of radium 224. It is still 3.64 days.

Few if any physicists believe that the many-worlds interpretation is valid. It is true that if one wants to reject Everett's theory, it is necessary to appeal to philosophical prejudice and say that the universe could not possibly be so complicated. But perhaps there is nothing wrong with doing this. After all, few individuals have ever accepted the philosophical theory of solipsism, even though it cannot be disproved either. The idea that I am the only thing that exists* is not one that I am able to take very seriously. Although there seems to be no way to disprove it, I doubt whether many philosophers would criticize me for rejecting it.

* Naturally, if you are a solipsist, you believe that *you* are the only thing that exists. In that case, *I* would not exist, and this book would be part of an illusion.

It so happens that there are a number of different alternatives to the Copenhagen interpretation of quantum mechanics. The many-worlds interpretation is just one. Even a physicist who believes implicitly in the ideas that were developed by Bohr would consider it worthwhile to explore these alternatives, even when they have implications that seem to be absurd. We would not be able to take the Copenhagen interpretation very seriously if we were unsure whether or not there were other ways of looking at quantum mechanics that were more plausible.

As it turns out, there are not any interpretations of quantum mechanics that really seem to be more reasonable. One of the reasons the majority of physicists still adhere to the Copenhagen interpretation is that the alternatives have been explored, and have been found to contain difficulties of various sorts. An entire book would be required to discuss this subject in detail. So I will limit myself to noting that the problem of the interpretation of quantum mechanics has been the object of intense research. Experiments have been performed (not all interpretations are as invulnerable to experiment as Everett's), mathematical theorems have been proved and the relative plausibility of different outlooks has been discussed. The motivation for this activity is obvious. One would have no idea what kinds of interpretation were possible if alternatives such as Everett's theory had not been looked at in detail.

It would be possible to argue that the many-worlds interpretation is not really a theory. After all, forming a theory and trying to figure out what interpretations the theory might allow are two different things. So perhaps it would be a good idea to take a look at an example of a real theory that is as farfetched as Everett's interpretation.

It really isn't necessary to look very far. Physics and cosmology are full of bizarre speculation these days. A good choice might be a hypothesis that has been proposed by Princeton University physicists Robert H. Dicke and P. J. E. Peebles. According to Dicke and Peebles, it is possible that a universe can reproduce itself.

It will be necessary to bring in some background material before we can look at the Dicke-Peebles theory in detail. In particular, it will be necessary to know something about *mini black holes* and about *virtual particles*.

A mini black hole is a black hole whose mass is much less than that of any star, or of any planet, for that matter. Mini black holes are too small to have been formed by stellar collapse. But they could have been created in the big bang that marked the creation of the universe.

According to currently accepted theory, the universe began in an enormous explosion that took place 18 billion years ago.* Initially, all of the matter in the universe was compressed into a very small volume. It is impossible to say how hot or how dense the universe was initially. Conditions were so extreme that we have no way of knowing how far we can extrapolate the known laws of physics. However, it is safe to say that at one point, there existed temperatures of billions of degrees, and that the density of matter was millions of times as great as that of terrestrial rock. Under such conditions, black holes could very well have been created in all conceivable sizes. In other words, numerous mini black holes might have been created early in the history of the universe.

If mini black holes exist, they should eventually disintegrate. Theory says that they should disappear after a certain period of time, and that intense bursts of radiation will be seen in their place. Although the term is not generally applied here, it would be possible to say that a mini black hole of a given mass has a certain specific half-life.

The time that it takes a mini black hole to disintegrate is related to its mass. The smaller ones would disappear first, the larger ones later. If the theory is correct, the only mini black holes that remain today should be those with masses greater than 10 billion tons.

Mini black holes would not be easy to detect. A black hole weighing 10 billion tons would have a diameter of about a tril-

---

* Recently, some doubt has been cast upon this figure. It is possible that it might soon be revised to something like 12 or 13 billion years.

lionth of a centimeter. It is possible that some of the gamma-ray bursts that have been detected by satellites are caused by the disintegration of mini black holes. But these bursts could very well have other causes. All we really know is that gamma rays originate somewhere in our galaxy and that some of them strike the earth. Since there are a number of different processes that could conceivably produce them, one must conclude that the existence of mini black holes has not yet been confirmed.

There is no theoretical lower limit on the size of a mini black hole. Although the smaller ones are most likely gone now, they could presumably still be created under the right conditions. It is conceivable that a mini black hole could be formed from a few subatomic particles. If one could just squeeze protons, neutrons or other particles close enough together, the gravitational forces might become significant. It is extreme compression of matter that creates a black hole, not a large amount of mass.

It is not likely that anyone will ever devise a way to make a black hole from a handful of particles, however. There are reasons to believe that it may not be possible to get particles that close together. If it could be done in principle, the problems would still be enormous. The energy required to accomplish such a feat would be so large as to be almost inconceivable. Even if we could build a particle accelerator the size of the entire universe, we might not be able to accomplish the task.

Black holes made from a few particles probably didn't even form in the big bang. It has been estimated that the first—and smallest—black holes to be created had a mass of about a hundred-thousandth of a gram. That doesn't sound like much. However, it is equivalent to the mass of about 10^{19} (10 billion-billion) protons or neutrons. It seems that we must conclude that black holes formed from just a few particles probably never did, and probably never will, exist.

Nevertheless, it is possible to speculate about the consequences that might follow if such black holes could be formed. This is precisely what Dicke and Peebles do. They make the assumption that few-particle black holes can form spontaneously, out of virtual particles.

Virtual particles are particles which are created in quantum fluctuations. They pop into existence for times of the order of 10^{-24} (one trillion-trillionth) second, and then disappear again. Their temporary existence is a consequence of the equivalence of matter and energy. According to Einstein's famous equation $E = mc^2$, matter can be transformed into energy and vice versa. If quantum uncertainties make a sufficient amount of energy available for a short period of time, virtual particles will be created. They disappear again almost at once because their existence creates an energy "debt" that has to be paid back almost immediately.

The concept of virtual particles sounds bizarre. After all, such particles seem to have only a ghostly kind of reality, and their lifetimes are so short that they cannot possibly be detected. And yet the theory of virtual particles has consequences that can be tested experimentally. The relevant experiments can be performed. The predictions of theory have been verified to a high degree of precision.

Theory implies that virtual particles are continually popping into and out of existence everywhere, even in empty space. The universe that modern physics describes is one in which there are "real" particles that are more or less stable, and these particles' virtual counterparts. If the theory is correct, the latter are much more numerous than the former. Every "real" particle is surrounded by a cloud of virtual particles. Additional virtual particles are located in interstellar and intergalactic space, where little matter is found.

This is the starting point of the Dicke-Peebles theory. The second step is an assumption that is entirely unwarranted. Dicke and Peebles assume that virtual particles do not always disappear within a small fraction of a second. They suggest that some of these particles form mini black holes before they can be destroyed. Under such circumstances, Dicke and Peebles say, the customary annihilation may not take place. The identities of the particles are destroyed when the black hole forms. They cannot disappear because they no longer exist.

There is no reason to believe that a process of this sort really takes place. But Dicke and Peebles realize this. They make this bizarre assumption because they want to see what the consequences would be. This, by the way, is a perfectly legitimate procedure. The idea that the velocity of light was independent of the motion of the source seemed bizarre when Einstein proposed it in 1905. However, as Einstein showed, this assumption had consequences that could be tested experimentally.

So even though the Dicke-Peebles theory already sounds outrageous, it might be worthwhile to follow their chain of reasoning just to see what happens at the end.

The next assumption that Dicke and Peebles make is outrageous too. They suggest that perhaps a mini black hole can avoid disintegration if it increases its mass. Then they conclude that a mini black hole that grew in this manner might eventually grow into a universe.

In the Dicke-Peebles theory, a mini black hole does not grow by swallowing up other matter. It grows because it is already a universe in miniature. Specifically, it is a tiny *oscillating universe*.

We can be fairly certain that *our* universe began with a big bang. Astronomers have not been able to determine what will happen to it in the future. Basically, there are two possibilities. Either it will go on expanding forever, or gravitational attraction between the galaxies will eventually cause the expansion to slow down, and then to stop. If the expansion eventually stops, gravity will continue to act, and a phase of contraction will begin. The contraction will continue until the entire universe is compressed together in a state which will be very similar to that which existed at the time of the big bang.

If this happens, no one can say what will occur next. However, since the 1930s, there has been speculation to the effect that the universe could somehow "bounce" and enter into a new phase of expansion. If this can happen, the universe might oscillate in an alternating series of expansions and contractions. What we call the big bang could be nothing more than the beginning of the latest expansionary phase.

Dicke and Peebles suggest that perhaps such oscillations can take place within a mini black hole. The particles of which the black hole is made are initially compressed together. After they come together in a singularity, they bounce back, expand and then come together in a singularity again. Every time one of these cycles take place, more mass is created.

If this theory is correct, then universes exist within universes like sets of Chinese boxes. After all, if our universe can reproduce itself in this manner, so can the universes that are created within it. And if this can happen, there is no reason why our universe cannot be a black hole within a larger one. It is theoretically possible for black holes to exist within black holes.

It is true that our universe does not look like a black hole. In particular, the matter within it has not collapsed into a singularity. But as we have seen, this might very well happen at some time billions of years in the future. Alternatively, the universe might be fated to go on expanding forever. In that case, there could exist mini universes within it, but it could not be a black hole inside a larger universe.

At this point, the reader may object that all this sounds much too outrageous to be believable. If he does, he may be right. I think it is safe to say that physicists do not take the Dicke-Peebles theory very seriously. It makes too many assumptions for which there is not any experimental or theoretical justification. Indeed, I would be surprised if I discovered that Dicke and Peebles believed in the theory themselves.

But the theory does have one attractive feature. It explains where our universe might have come from.

It had previously been thought that there were only two possibilities. Either the universe was created in the big bang, or it had always existed, going through an endless series of cycles. The Dicke-Peebles theory suggests that there is also a third possibility. Our universe could have been created out of nothing inside a universe that already existed.

Could the theory be true? Probably not. It depends upon too many unlikely assumptions. But it is not a crackpot theory. Dicke and Peebles do not ignore the known laws of physics. On the contrary, they try to see what possibilities these laws allow.

When matter collapses into a singularity, even the laws of general relativity break down. The conditions become so extreme that Einstein's theory can no longer be valid. This prevents us from knowing what goes on in the singularity in a black hole. And it prevents us from extrapolating back to the beginning of the big bang. Matter was very compressed at the beginning of the universe too. In fact, physicists sometimes speak of the big bang as the "initial singularity."

When one does not know what happens in a given situation, it is possible to assume anything, just to see what the implications are. This is what Dicke and Peebles were doing when they propounded their theory. They were not really attempting to discover what was true; they were trying to find out what might possibly be true.

This type of theorizing is not uncommon in contemporary physics. If science is to find out what the universe is like, it must first discover what is possible. So many fantastic discoveries have been made in the past, and so many commonsense ideas have been overthrown that one must be willing to consider any idea— no matter how outrageous it may seem—as long as one doesn't leave the framework of science entirely when one does so.

One might think that this is a futile kind of activity. But it is not. If the Dicke-Peebles theory eventually turns out to be wrong —as it probably will—it may still be useful to theoretical physicists. It may very well turn out that some assumption made by Dicke and Peebles will be useful in a theory of the future.

No one expects that everything brought back by those who explore the frontiers will be useful. But that does not imply that imaginative theoretical explorations are of no worth. The Peebles-Dicke theory is a variety of science fiction, to be sure. But sometimes science fiction has its uses.

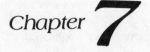

CONNECTIONS

RECENTLY IT HAS BECOME FASHIONABLE to perceive analogies between the theories of modern physics and the insights of Eastern mysticism. One author, physicist Fritjof Capra, goes so far as to claim that "modern physics leads us to a view of the world which is very similar to the views held by mystics of all ages and traditions."

This is really not a new idea, however. It was stated more than thirty years ago by the British novelist Lawrence Durrell. In his book *A Key to Modern British Poetry*, published in 1952, Durrell claimed that it was possible to see parallels not only between science and mysticism, but also between mysticism and modern poetry. Scientists and mystics alike, Durrell said, had found it necessary to "marry opposites in order to arrive at a picture of reality." Since similar trends could be perceived in literature, it was clear that poets, mystics and scientists were "converging upon a single objective."

As far as I can tell, Durrell's ideas were not taken very seriously. In any case, they were soon forgotten. But when Capra published *The Tao of Physics* in 1975, the response was entirely different. As I write this, Capra's book has sold something like a half-million copies, and has provided the inspiration for numerous other books and articles dealing with parallels between modern science and Eastern thought.

There are probably a number of reasons why Capra's book should have been so much more influential than Durrell's. In the first place, Durrell really did not understand modern science very well. When he attempted to elaborate upon theories of physics, he sometimes managed to make three or four mistakes in a single paragraph. Furthermore, he exhibited a tendency to overstate his case. He maintained, for example, that a relativistic conception of time could be found in the writings of the Irish novelist James Joyce and in the work of the French author Marcel Proust. To readers who doubted that Joyce and Proust really understood relativity any better than Durrell did, this must have seemed a dubious claim. Although *A Key to Modern British Poetry* contains some very penetrating analyses of the work of modern British poets, one comes away with the feeling that whenever Durrell begins to talk about science, his enthusiasm outruns his understanding.

Capra, on the other hand, is a physicist who has a knack for writing lucid explanations of the concepts of modern physics. He effectively conveys a feeling for the nature of quantum-mechanical concepts to his readers, and does a good job of explaining such ideas as relativistic space-time. One concludes that he understands the things he is talking about, whether he is discussing mysticism or science. This is really not very surprising. Since Capra practices T'ai Chi—Taoist meditation—he presumably possesses insights into mystical practice that Durrell lacked.

Another factor that undoubtedly contributed to the success of Capra's book is the fact that many more Westerners were interested in Eastern thought in 1975 than in 1952. During the 1960s, mystical philosophies became a preoccupation of millions of people. These individuals must have been happy to find an author who assured them that the truths of mysticism were confirmed by the findings of modern physics.

However, we must bear in mind that a book or an idea can be popular and influential and yet contain mistaken ideas. If the idea that there are parallels between mysticism and physics has become fashionable, it does not necessarily follow that it is true. Indeed, I would tend to think that it is the fashionable ideas that should be subjected to the most severe scrutiny. Ideas that are beguiling are the very ones that are most likely to lead us astray.

The idea that there are parallels between physics and mysticism is not one that can easily be dismissed. After all, quantum mechanics does sometimes seem to have its paradoxical aspects, and paradox is a hallmark of mystical thought. Over and over again, mystics have insisted that the Absolute can be described only in terms that seem contradictory. Ordinary language, they say, is not adequate to describe the reality that is hidden behind the world of appearances. The individual who attempts to speak of it must inevitably make use of concepts that do not seem to obey the rules of ordinary logic.

In spite of all this, one must consider the idea that parallels exist to be a very surprising claim. After all, science and mysticism are two very different kinds of activity. The mystic seeks illumination within himself, while the physicist attempts to understand the natural phenomena that he sees around him. Mysticism is a technique for attaining enlightenment. In science, the spiritual state of the individual scientist should not make any difference.

However, the topic of science and mysticism is a complicated one. There exists a certain amount of circumstantial evidence that science and mysticism might be similar in one way or another. Specifically, the physicists who created quantum mechanics were often intrigued by Eastern outlooks. They did not claim that physics and mysticism led to equivalent views of reality. Nevertheless, they apparently did sometimes see something in the mystical world view that reminded them of their attempts to understand the nature of the quantum world. Bohr included the Chinese yin-yang symbol in the coat of arms that he chose when he was knighted. Schrödinger developed a profound interest in Eastern thought. In his influential book *What Is Life?* he commented upon the implications for biology of quantum mechanics and of Eastern philosophy. He claimed, for example, that only Eastern thought could adequately explain the nature of consciousness. And of course, Einstein seems to have been something of a mystic himself. Although he seems not to have been interested in any of the Eastern religions, he apparently did think of the universe as something that partook of the divine.

I think that one should conclude, however, that none of this really proves anything. Anecdotal evidence of this sort is never very persuasive. Generally, Capra and the other authors who write on physics and Eastern tradition don't even bother to mention it.

After all, if Bohr and Schrödinger exhibited some interest in mystical concepts, that should not surprise us. In their scientific work, they had had to learn to throw aside commonsense ideas in order to understand the nature of events in the subatomic world. It is only natural that they should have been intrigued to discover that the mystics had thrown aside commonsense concepts also. The fact that they exhibited this interest proves nothing, any more than a chess grandmaster's interest in Chinese food would demonstrate that there was a connection between chess and Oriental cuisine.

Bohr does seem to have seen a parallel between the concept of yin and yang and his own doctrine of complementarity. But we should probably not attach very much significance to this fact. Bohr seems to have believed that complementarity was a principle that could be applied in many fields outside of physics, including psychology, biology and anthropology.

So perhaps it would be best to examine the arguments given by Capra and by other authors for the unity of mysticism and physics. The idea that is emphasized most often is the one that both physics and mysticism illuminate: "the unity and interrelation of all things," as Capra puts it. The universe of modern physics, he says, is an interconnected whole. This, he adds, is precisely what is perceived by the Eastern sages.

When Capra claims that modern physics reveals that there exist interrelations between the fields, forces and particles that make up the physical universe, one must agree with him. After all, Einstein showed that space and time are interrelated. He showed, furthermore, that gravitational interactions give the entire universe a particular "shape" and structure. It is the presence of gravitational masses which creates the curved geometry of space-time.

Capra also makes a valid point when he talks about the interrelations between the various subatomic particles, pointing out that such particles as the proton and the neutron do not even maintain their identities. A proton, for example, can emit a particle called a *pi meson*, or *pion.** If a proton emits a pion that is positively charged,† the proton will be transformed into a neutron. If a neutron absorbs a positively charged pion, it will be changed into a proton. The same thing will happen if the neutron emits a negative pion instead of absorbing a positive one.

This implies that the neutrons and protons in a nucleus are constantly exchanging identities. In fact, they can be thought of as two different manifestations of a single particle, the *nucleon*. Physicists often regard the neutron and proton as nothing more than two different *isotopic-spin* states of the same particle. Isotopic spin, by the way, has nothing to do with ordinary spin; it is an abstract mathematical concept.

There are yet other interrelations. A pion can transform itself into two nucleons, for example. The nucleons may then come together to form the original meson again, or they may go their respective ways and participate in yet other reactions. If a meson decays into two nucleons, things are a bit more complicated. Nucleons are never created alone, but only in particle-antiparticle pairs.†† For example, a pion can give birth to a neutron and an *antiproton* (an antiparticle which is similar to the proton, but

---

* Mesons (the pion is one of many kinds) are particles that are associated with the forces that bind protons and neutrons together in an atomic nucleus. They are a kind of nuclear "glue."

† There are three kinds of pions: positively charged, negatively charged and neutral.

†† At least, no exceptions to this rule have been observed. Later, I will discuss the possibility that there might be rare exceptions.

which has a negative, rather than a positive, charge). Or it may transform itself into a proton and an *antineutron,* or proton and antiproton, or neutron and antineutron. The only requirements are that electric charge be conserved, and that nucleons and anti-nucleons be created it pairs. A neutral pion cannot change into a proton and an antineutron because a positive charge cannot be created out of nothing. It can't change into two neutrons; in that case particles and antiparticles would not balance each other out.

It all sounds very complex, and it is. The subatomic world is a sea of ferment. Particles are constantly interacting, exchanging identities and reacting to the influence of everything around them.

Capra says little about *quarks,* the theoretical particles from which nucleons and mesons are formed. He is an advocate of *bootstrap theory,* which was considered to be a major theoretical alternative to quark theory at one time. Bootstrap theory can be roughly described by saying that it was a theory which denied that particles had any "fundamental" constituents. Everything was made up of everything else. Pions were made of nucleons, and nucleons were made of pions. There was, so the theory said, no bottom level.

Bootstrap theory is not very fashionable nowadays. But this does not really lessen the force of Capra's argument. One can easily apply the reasoning that Capra uses to quarks also. There is nothing about his outlook that would restrict its application to mesons and nucleons alone.

According to currently accepted theory, neutrons and protons are each made up of three quarks. Mesons are made up of quark pairs. There also exist particles called *gluons* which bind the quarks together in pairs or triplets. The role of the gluons is exactly analogous to that of the mesons which "glue" neutrons and protons together in a nucleus. Quarks and gluons participate in reactions similar to those which the nucleons and pions undergo. Whether one is looking at nucleons or quarks, the same web of interrelationships appears.

Capra claims that these webs of interrelationships constitute an aspect of the "unity of all things" that is spoken of by mystics. In my opinion, it is at this point that his argument breaks down. It seems to me that he shows only that an analogy can be made. He does not show that one kind of interrelationship has anything to do with the other.

As we all know, analogies prove nothing. In fact, they are sometimes quite misleading. For example, astronomy and astrology both speak of interrelationships between the various planetary motions. But the success of the former does not demonstrate the validity of the latter. Whenever one encounters an argument from analogy, it would be well to remember Lewis Carroll's question "Why is a raven like a writing desk?" Carroll's point was that one can always find a way to compare anything to anything else, no matter how dissimilar they may be.

I find Capra's argument unconvincing from a scientific point of view, and I would not think highly of a mystic who accepted it either. It really doesn't seem to have much validity when viewed in the context of a mystical interpretation of reality.

According to the mystics, reality can be divided into a number of different levels. The actual number varies from one mystical tradition to another, but the world is generally subdivided into constituents that can be more or less accurately labeled *matter*, *mind* and *spirit*. In every mystical tradition, matter is considered to be the lowest and least significant of the three. Mystics tend not to be very much interested in events on this level, and they generally repudiate the idea that an understanding of the subtleties of matter can have any significance for insight into the Absolute. Or as Ken Wilber, a writer on mystical traditions, puts it, "interrelated shadows are not the Light." In other words, interrelations in the observable world can tell us nothing about the "higher reality" that mystics claim to see.

I find arguments advanced by some of the other authors who write about "mystical physics" to be even less convincing. In particular, I don't think one should be swayed by the argument that relativity, quantum mechanics and mysticism must all be alike because they all contain paradoxes.

It is often pointed out that the utterances of the sages are frequently paradoxical in nature. It is said that since modern physics is paradoxical too, the mystics and the physicists must really be talking about the same thing.

It seems to me that this argument contains a huge fallacy. Most of the "paradoxes" of modern physics are really not that at all. An idea that is logically paradoxical is one thing. A concept that does not conform to commonsense notions is quite another. Although the latter are quite common in modern science, the former are not. A scientific theory that engendered real paradoxes would soon have to be given up as inconsistent.

The discussion of paradoxes played an important role in the revolution in physics that took place in the early years of the twentieth century. However, these were generally paradoxes found in the old, classical theories. Planck formulated his quantum hypothesis because it seemed to be the only way to resolve the paradoxes in the classical treatment of blackbody radiation. Einstein's discovery of special relativity resulted, in part, from a consideration of the paradoxes that resulted when one tried to imagine what would happen when an object moved at the velocity of light. The wave/particle duality was thought to be paradoxical for a time. But the paradox disappeared when physicists realized that it was possible to give up the commonsense notion that an object had to be either a wave *or* a particle. Once they understood that a distinction which was valid in the world of macroscopic objects need not be carried over to the world of quantum events, the difficulty vanished. A truly paradoxical description of physical reality would not be possible. If scientific theories are to explain natural phenomena, they must be capable of giving unambiguous predictions.

Are physics and mysticism related? Very likely not, except in the sense that they are both creations of human minds. If the Tao really exists, it seems that one must seek it by practicing the traditional methods for attaining enlightenment, rather than by studying particle physics, relativity and quantum mechanics. In spite of all the arguments advanced in the books that have been written on the subject of physics and mysticism, it seems to me that religion is still religion and physics is still physics.

Sometimes I wonder whether the authors who speak of mysticism and physics are not engaged in an illegitimate attempt to use the success of physics to support the claims of mysticism. After all, it is not all that obvious to the skeptic that mystical insights have much validity. Insights can presumably be mistaken whether they are of the scientific or of the religious variety. It is true that mystical experiences seem always to be accompanied by a feeling of certainty. However, this particular emotion seems to be an insufficient guarantee of truth. As we have seen, Kepler experienced it when he "discovered" that the solar system was built of geometrical solids.

In fact, I am not sure that there is really very much difference between attempts to use physics to justify mysticism and Velikovsky's attempt to use a "scientific" theory to support a fundamentalist interpretation of the Old Testament. However many books are written on such subjects, we are probably safe in assuming that the Tao has not been detected in the laboratory.

If there is probably not any connection between physics and mysticism, there could still be meaningful parallels between the sciences and the arts. After all, artistic creation and scientific discovery are both products of the human imagination. Physicists and poets alike have spoken of experiencing sudden insights, and the inspiration that comes to a painter may not be much different from that which is experienced by a mathematician.

For example, the nineteenth-century German mathematician and astronomer Karl Friedrich Gauss once reported that after he had been unsuccessfully struggling with a mathematical problem for years, the solution came suddenly, like a "flash of lightning." After the problem had been solved, Gauss found that he could not say exactly what it was that had made his success possible. The French mathematician Henri Poincaré experienced something similar. It seems that Poincaré had been wrestling with certain mathematical quantities called fuchsian functions. The solution to one of his problems appeared suddenly as he put his foot on the step of a bus while on a geological excursion. According to Poincaré, he had not been thinking about the problem at all; the incidents of travel had caused him to forget his mathematical work entirely.

When Poincaré experienced his insight, he did not have time to verify the result. So he took his seat on the bus and resumed his conversation with another traveler. Only after he returned home did he have the chance to check the result that had come to him. But when he did check it, it turned out to be correct.

One cannot help being reminded of what Mozart said about his method of composition. Mozart often found that musical ideas would crowd into his mind as he was taking a walk after dinner. In a celebrated letter, Mozart reported that this required no act of will; he simply could not stop the ideas from coming. An entire work would sometimes seem to appear in his mind before he had written down a note.

The experiences of the French poet Paul Valéry were also similar to those of Gauss, Poincaré and Mozart. Valéry spoke of poetic inspiration as "a gleam of light, not so much illuminating as dazzling." Although the poetic fragments that came to him did not "illuminate the whole mind," they did sensitize it, and point to additional ideas that could be incorporated into his work.

One of the most extraordinary accounts of creative discovery, however, was given not by a poet or a composer but by a chemist. In 1865, the German chemist Friedrich August Kekulé von Stradonitz had been trying to understand the structure of the substance benzene. Now, it so happens that benzene has a ring-like structure. The component atoms of the benzene molecule link up in a hexagonal pattern. Before 1865, chemists had not dreamed that such a structure could be possible. Consequently, until 1865, Kekulé had been operating on the assumption that the atoms of the benzene molecule were arranged in some kind of linear chain.

Kekulé continued to struggle with the problem with no success. And then, one afternoon, he sat in a chair in front of a fire and dozed off. This is his description of what happened next:

Again the atoms were gamboling before my eyes. This time the smaller groups kept modestly in the background. My mental eye, rendered more acute by repeated visions of this kind, could now distinguish larger structures, of manifold conformation; long rows, sometimes more closely fitted together; all twining and twisting in snakelike motion. But look! What was that? One of the snakes had seized hold of its own tail, and the form whirled mockingly before my eyes. As if by a flash of lightning I awoke.

Kekulé had discovered the benzene ring.

Kekulé's vision is somewhat reminiscent of one experienced by the English poet Samuel Taylor Coleridge, which led to the writing of the poem *Kubla Khan*. But Coleridge's dream seems to have been induced by opium, while Kekulé's insight was aided by nothing but a warm fire. Of the two, the latter may have been the more revolutionary (but not necessarily the more significant; scientific and artistic significance can hardly be compared). When Coleridge published his poem, he presented it as a "psychological curiosity" that he was allowing to be printed at the urging of his friend Lord Byron. Kekulé's discovery that ring structures were possible was universally recognized as something extremely important. In the end, it revolutionized organic chemistry.

Although scientific and artistic inspiration seem to be very much alike, it is not so easy to find similarities between scientific theories and finished artistic works. Durrell's notion that a relativistic conception of time can be discerned in such works as Joyce's *Ulysses* is not very convincing. Neither is the idea, suggested by the Austrian art historian Sigfried Giedion, that there are parallels between modern physics and trends in early-twentieth-century painting.

When Giedion gave the Charles Eliot Norton Lectures at Harvard University in 1938 and 1939, he suggested that there were parallels between Einstein's relativity theories and the Cubist and Futurist schools in art. According to Giedion, there had been a separation between "thought and feeling," between the sciences and the arts, during the nineteenth century. These separate currents, Giedion said in his lectures (and in his book *Space, Time and Architecture*), had begun to come together again during the early years of the twentieth century. As a result, identical themes had begun to find expression in painting and physics alike.

This idea seems somewhat less plausible today than it did forty-five years ago. It would be difficult to find scientific equivalents for such recent movements as Abstract Expressionism and Pop Art. And as Giedion admitted, there do not seem to be any scientific parallels to the Romantic movement or to such schools as Impressionism or Expressionism. It is hard to believe that science and art somehow managed to parallel each other during the first decade or two of the twentieth century when they did not do so before that time, and have not done so since.

There may have been something about the early years of the century that encouraged artists and scientists alike to try out new ideas. The period was characterized by the introduction of audacious new concepts into physics, and by experimentation with novel artistic and literary techniques.

However, to suggest that the art and science of the period paralleled each other in more fundamental ways may be carrying the analogy too far. Maintaining a skeptical attitude might be wisest. The idea that there are parallels between science and the arts does not seem any more convincing than the idea that there are parallels between science and mysticism.

Artistic creation and scientific discovery are two different kinds of creative activity. For one thing, the language of science must be precise, while that of art makes use of ambiguity. An idea in physics must be given a precise mathematical formulation if its implications are to be worked out in detail, and if experimental tests are to be devised. Mathematics in one sense is nothing but a kind of language that allows scientific statements to be formulated with greater precision than is possible in English (or in Japanese, or in Russian, or in any other language that a physicist might speak).

On the other hand, the impact of a poem or a musical composition or a painting often depends upon a kind of built-in ambiguity. This ambiguity helps the reader's or the viewer's imagination to come into play. One reason the paintings of Norman Rockwell seem less than great art is that their messages are too obvious. With the portraits of Rembrandt, however, the longer we look at them, the more deeply we experience their emotional nuances. Similarly, pulp literature typically makes bald statements concerning its characters' feelings. Great literature requires that the reader make an effort to understand its characters through feelings of empathy.

Another obvious difference is that science progresses, while art does not. Although an artist may choose either to elaborate upon the themes used by the artists of a previous generation or to react against them, he really does not discover or express anything about human experience that was not known to his predecessors. Scientific discovery, on the other hand, does build upon the discoveries of past generations of scientists. With each passing year, science understands the universe a little better.

This is the reason that great works of art seem never to lose their appeal, while great scientific discoveries become entombed in textbooks. Shakespeare is read by greater numbers of people in each succeeding generation. But few physicists ever bother to read the original papers of Einstein. They know that the explanations of relativity which appear in textbooks contain all the significant ideas, while avoiding the obscurities that were occasionally present in Einstein's writings.

Finally, style is all-important in art. We value a novel or a sculpture not only for what its creator says, but also for how he says it. If this were not the case, no one would bother to read novels; after all, a nonfiction essay can be used to express an idea more succinctly.

An artist may spend a lifetime elaborating upon stylistic innovations. On the other hand, although scientific style does exist, it is relatively less important. A creative scientist is remembered for what he said, not for how he said it.

In spite of all these differences, there seem to be certain odd, inexplicable connections between the arts and the sciences. For example, there seem to be respects in which mathematics and music are alike. Although I can think of no great mathematicians who were also noted musicians, it has been observed that musical and mathematical ability are often found in the same individuals. Of all scientists, it is the mathematicians who most value esthetic qualities in their work. One commonly hears mathematicians speak of "elegant" proofs. And of course, the quality of elegance is also valued in a musical performance.

The consonance or dissonance of a musical chord is a quality that can be described mathematically. Mathematical ratios are also used in the construction of musical scales. But these factors may not have much to do with the puzzling analogies between mathematics and music. After all, a musician does not have to know what these ratios are. It is necessary only that he have a good "ear."

Music is the most abstract of the arts, while mathematics is the most abstract of the sciences. It is more likely that the analogies have something to do with this. However, this observation does not explain why, every now and then, a six-year-old child should learn to play the violin with an amazing mastery, and why there should occasionally be nine-year-old mathematical geniuses. The abstractness doesn't explain why this should be the case, when children do not make great discoveries in physics, or create paintings that have any special significance.

It seems that all that one can say is that there appears to be a vague, puzzling similarity between musical and mathematical endeavors. There seems to be a sense in which mathematicians and musicians are engaged in doing something similar. But it does not seem to be possible to say exactly what this "something similar" is.

There may also be a relationship of sorts, a somewhat more tenuous one, between physics and poetry. Admittedly, the evidence is less compelling in this case. However, there seems to be enough to at least suggest that there is a sense in which the two endeavors are alike.

Both physics and lyric poetry* seem to be the province of young men (and women). Einstein was twenty-six years old when he published his special theory of relativity, and Newton was about the same age when he discovered his law of gravitation. Both men experienced a falling-off in scientific productivity as they grew older. Einstein seems to have been over the hill by the time he was forty, and Newton devoted his later years to studies of theology and alchemy, and to his duties as Governor of the Royal Mint. The same pattern continues to hold today. Nobel Prizes in physics are frequently awarded for discoveries that were originally presented in doctoral dissertations, and important advances are rarely made by physicists who have reached middle age.

Something very similar can be observed among poets. Although great poems have been written by individuals who were fifty or older, the poetic imagination appears to be most active in youth. Novelists typically find that they are just beginning to hit their stride when they reach the age of forty, but a forty-year-old poet will often discover that the bulk of his best work has already been written.

Some poets begin to produce significant work while they are still adolescents. Although the Welsh poet Dylan Thomas died before he was forty, he completed only about one poem a year during the last decade of his life. On the other hand, he had written drafts of half the pieces that were to appear in his *Collected Poems* before he was twenty. The Austrian poet Hugo von Hofmannsthal (who is perhaps best remembered today for the librettos he wrote for the composer Richard Strauss) became famous for the poems he wrote while he was still in his teens. The French poet Arthur Rimbaud wrote no poems at all after he was nineteen. Of course, there are exceptions. The Irish poet William Butler Yeats and the English poet and novelist Thomas Hardy are best remembered for poetry that they wrote during the latter parts of their lives. However, Yeats and Hardy are not

* I distinguish here between lyric poetry, such as that written by Keats, and the epic or dramatic poetry of such authors as Homer and Shakespeare.

typical. The bulk of the work that is included in any standard anthology of poetry consists of poems written in the authors' youth.

There are no eighteen-year-old physicists. Poetry and physics differ in that a great deal of formal education is necessary if one wants to work in the latter field. It is necessary to become familiar with the discoveries of the past before one can hope to do original work. However, in other respects the careers of poets and physicists exhibit a certain resemblance to each other, and to those of professional athletes as well. If a thirty-five-year-old quarterback can be said to be "aging," the same remark could reasonably be made of a poet or physicist of the same age. Just as the professional football player finds it difficult to continue to compete when he begins to lose the bodily vigor of youth, the poet and the physicist find it hard to keep up the pace of their work as they grow older.

It appears that one must have a fresh, unencumbered imagination in order to write poetry prolifically, or to make significant discoveries in physics. Possibly the mind begins to run in set patterns as one enters the fourth and fifth decades of life. Or perhaps it is necessary to have a childlike sense of wonder if one is to accomplish anything significant in either poetry or physics.

When Beethoven was forty-seven, he remarked to a friend, "*Now* I know how to compose." It would be easy for a scientist or an artist in any of a number of different fields to make this kind of remark. Many composers, and painters and sculptors as well, are at a peak of creativity when they are forty-seven. Forty-seven is young for a philosopher, and it is an age at which important discoveries are made in many scientific fields. But the forty-seven-year-old poet characteristically finds that his output has declined dramatically, and a forty-seven-year-old physicist can often do nothing more than elaborate upon the discoveries of his youth.

There is much about creativity that is still mysterious. Most often it is as mysterious to the creative individual as it is to anyone else. Poets frequently point out that they find it difficult to tell where their ideas come from. Some of them have reported experiencing the feeling that they "do not really write" their poems. In many cases, inspiration suddenly "comes" to them, as if out of nowhere.

Scientists—both physicists and those who work in other fields—frequently have the same feeling. Poincaré's tale of the sudden insight that he experienced as he put his foot on the step of a bus is one that would be familiar to creative individuals in many different fields.

At one time, it was believed that there was something divine about poetic inspiration. A poet in Greek or Roman times habitually began a work with an invocation to the Muse. It seemed only natural to believe that he would not be able to write if she deserted him. In Norse mythology, creativity—the mead of poetry —was an inadvertent gift of the gods.

Although the topic is not so often discussed, it is obvious that some scientists have also felt that they were peering into the divine. Einstein had the feeling so often that he spoke of the natural order of the universe as though it had a divine nature. Kepler's intoxication over his mistaken discovery of planetary orbits and perfect solids was promulgated with the fervor of an Old Testament prophet. In this case, the fact that the insight was really a blunder seems not to have made any difference; when it came, it was as convincing as true insights are, and its origin was just as mysterious.

We must conclude that creativity in poetry and creativity in physics are alike in some sense, and that scientific and artistic creativity are alike in some more general way. It is extraordinarily difficult to add anything more. The similarity is intriguing, but recognizing that it exists does not advance our understanding of scientific discovery as much as one would hope.

So perhaps it would be best to leave the topic for now and to look at yet another kind of connection. What I have in mind is not the possible connections between science and mysticism, or poetry or anything else, but rather the webs of connections that exist within science itself. In my view, these interconnections within science may be the most significant ones of all.

I think that perhaps I can best explain what these connections are by giving an example of something that most of us would not consider very scientific: the theory that the earth is flat.

Contrary to what one might think, it is really not so difficult to put together a theory which explains observed phenomena with a flat-earth hypothesis. After all, it is possible to explain anything if one only makes enough hypotheses. All that the flat-earth theorist really has to do is invent an assumption to explain every phenomenon that he can think of, and then add another assumption every time an objection is made to the resulting theory.

For example, the disappearance of ships over the horizon can be explained by making some sort of assumption about the refraction of light in the air. Moon landings and photographs of a spherical earth taken from the moon could be explained by the paranoid hypothesis that it was all staged. There is no way that one can prove that the earth is a sphere to someone who is willing to add another *ad hoc* assumption to his theory every time he encounters a situation in which it does not work.

A theory of this sort would not be a very appealing one. It would seem much too complicated to be a true picture of reality. In fact, it would be in much the same class as the Ptolemaic theory of the solar system, which was rejected by Copernicus, by Kepler and by Galileo because it seemed to contain too many *ad hoc* assumptions.

Elegance is one of the hallmarks of a believable scientific theory. If a theory can successfully explain a wide variety of phenomena by appealing to a relatively small number of assumptions, it will often be accepted even before it has been confirmed experimentally. Newton's law of gravitation provides a good example of this. By making the simple assumption that gravitational forces were described by an inverse-square law, Newton was able to explain the motions of the planets, the trajectories of projectiles in the earth's atmosphere and the nature of the tides. Newton's contemporaries accepted the theory relatively quickly because they found it difficult to believe that a theory which explained so much by assuming so little could possibly be false.

Successful theories sometimes produce unexpected results, and become even more wide-ranging than they appeared to be at first. For example, Einstein's original paper on special relativity made no mention of the famous equation $E = mc^2$. It was only after his paper had been sent off for publication that Einstein realized that his theory implied anything about the equivalence of mass and energy. So he supplemented his theory by writing a second paper which dealt specifically with the mass/energy formula. This paper was published later the same year.

Sometimes the wide-ranging applicability of a theory causes scientists to give serious consideration to ideas that they did not previously take very seriously. When the Russian-born American physicist George Gamow published his big-bang theory of the origin of the universe in 1948, there were many physicists who declined to take Gamow's ideas very seriously. They did not really believe that it was possible to say anything about events that had presumably taken place billions of years ago. Some of them regarded the hypothesis that the universe originated in an enormous fireball as pure fantasy.

The big-bang theory is considered to be anything but fantasy today. The reason that it is now generally accepted is that it explains a number of diverse phenomena, and establishes the existence of connections that had not previously been preceived. Not only does the big-bang theory explain the present expansion of the universe; it also tells us why the universe should be approximately 25 percent helium, and why radio waves of certain wavelengths should be falling on the earth from every direction of the sky.

The radio waves, known technically as the *cosmic microwave radiation background*, were discovered by Bell Laboratories scientists Arno Penzias and Robert Wilson in 1964. Penzias and Wilson received the Nobel Prize for their work in 1978. By that time, it was generally agreed that the radio waves that they had discovered were degraded light from the primeval fireball in which the universe was created.

Evidence for this conclusion is provided by the fact that the radio waves resemble those which would be emitted by a black-body 3 degrees above absolute zero ($-270°$ C—often written 3 K; K stands for Kelvins, or degrees above absolute zero). This is exactly the kind of radiation that Gamow's big-bang theory predicts. During the early stages of the fireball, temperatures reached billions of degrees, and the radiation that was emitted was accordingly intense. In the billions of years that have passed since that time, the average radiation temperature of the universe has dropped dramatically, so that only low-energy radio waves remain. One way to picture the transformation of high-energy radiation into low-energy radiation would be to imagine the wavelengths expanding in order to adjust to a universe that was rapidly growing larger. Since long wavelengths are associated with lower energy, light, X rays and gamma rays would all eventually transform themselves into radio waves.

Evidence for the big-bang theory is also provided by the fact that the universe contains approximately 25 percent helium. Whenever the helium content of stars, galaxies or interstellar clouds of gas is measured, this is the figure that is obtained. The results never vary by more than a few percentage points one way or the other.

If one does not assume a big-bang origin for the universe, it is difficult to explain where this helium comes from. Nuclear reactions that take place in stars do convert hydrogen into helium, but this conversion takes place at a very slow rate. It is estimated that only a small fraction of the helium present in our own galaxy could have been formed in this way; the galaxy has not been around long enough to produce helium in significant amounts. The greater part of the helium must therefore already have been present when the galaxy formed.

The big-bang theory solves the problem by demonstrating that the helium could have been created by nuclear reactions which took place in the primeval fireball. Moreover, detailed calculations indicate that exactly the right amount should have been created in this manner. The theory predicts the concentration of helium that we see. Furthermore, the big-bang theory also explains why deuterium, an isotope of hydrogen that is *not* made in stars, should be present in concentrations of 20 to 30 parts per million. The deuterium was created in the big bang too.

Naturally, it would be possible to invent alternative explanations for the cosmic background radiation. Any reasonably inventive astrophysicist could presumably dream up hypothetical mechanisms for the production of helium and deuterium. However, it would be necessary to use *ad hoc* assumptions that would explain each phenomenon separately. Instead of one relatively simple theory, we would have two or three complicated ones.

The big-bang theory is a successful theory because it ties so much together. It explains the expansion of the universe, the existence of the radio waves which look as though they had come from a blackbody and the helium and deuterium concentrations that appear to be about the same everywhere. The big-bang theory shows that there are important connections between these apparently unrelated phenomena, and it accomplishes this without making a lot of complicated assumptions.

It is sometimes said that a theory which exhibits qualities like those of the big-bang theory possesses the characteristic of *simplicity*. But the use of this term can be rather misleading. Many supposedly "simple" theories are actually quite complicated. For example, the equations of general relativity are so complex that they can be solved only in special cases. Even Newton's law of gravitation can present great mathematical difficulties. For example, no one has ever found an exact solution for the Newtonian *three-body problem*. If one takes three gravitating bodies—say, the sun, the earth and the moon—it does not seem possible to find a mathematical formula that will describe all the intricacies of their motion. In practice, one could solve the problem to as high a degree of approximation as one wanted, provided that a computer was available. However, as far as we can tell, only approximate solutions seem to be possible.

Sometimes I think that the word "parsimonious" would be the best one for describing a theory which obtains far-reaching results while making only a small number of uncomplicated assumptions. It seems to me that the important thing is not so much the simplicity of the initial assumptions, but the ability of a theory to find unexpected connections, to explain a wide variety of phenomena without the introduction of additional hypotheses.

Successful theories explain so much with so little that there seems to be something almost miraculous about them. Copernicus assumed only that the earth moved around the sun, and this was sufficient to allow Kepler to work out the laws of planetary motion. When Gamow worked out the big-bang theory, the only assumption he made was that if the universe was expanding now, it must originally have been in a compressed state. Einstein's two theories of relativity are not based on much more than the assumption that it should be possible to write down the laws of physics in such a way that they are the same for an observer in any state of motion. Each of these assumptions was deceptively simple, and yet each had so many consequences that it profoundly altered our conception of the universe.

But how do we know that theories must be simple (or parsimonious) if they are to have any chance of being true? The answer to this question should be fairly obvious: we don't. The postulate of simplicity is a philosophical assumption. The astronomers who looked at the Ptolemaic theory and said, "Things cannot be as complicated as all that!" were not expressing a scientific objection, but rather a philosophical one.

We cannot be certain that nature is really parsimonious any more than we can be sure that eccentric philosophical theories like solipsism are not valid. We assume that nature is based on simple laws because we have no choice. If this assumption were not valid, scientific knowledge would not be possible. We would not know what to believe, because we would not be able to choose between complicated flat-earth theories and uncomplicated theories of a spherical globe.

The philosophical assumption of simplicity, or parsimony, that underlies scientific theory can also be interpreted as an esthetic principle. A theory that achieves its ends with the utmost economy of means is frequently described as "beautiful." On the other hand, a theory that is too complicated to be plausible can be aptly described as "ugly."

When John Keats wrote his famous poem "Ode on a Grecian Urn," he found that the figures painted on the urn seemed to be saying to him, "Beauty is truth, truth beauty." This is the kind of statement that one would expect a poet to make. It is somewhat surprising to find that it applies to science as well.

REASSEMBLING
THE UNIVERSE

I HAVE USED the term "dismantling the universe" as a metaphor for scientific discovery. Scientific revolutions come about when scientists perceive that there are problems with the picture of the universe that they have been using. They respond by dismantling this conceptual universe, replacing it with something else.

The Ptolemaic system was dismantled by Copernicus, by Kepler and by Galileo because they felt that this theory contained too many *ad hoc* assumptions to be reasonable. They rebelled at the idea of adding yet more epicycles, eccentrics and equants in order to reconcile it with observational data. Einstein proposed his special theory of relativity because he felt that there were inconsistencies in the world view of classical physics. Specifically, the laws of physics did not always look the same to observers in different states of motion. In any reasonable kind of universe, Einstein believed, this would not be the case.

Surprisingly, new experimental findings do not have much to do with the dismantling process. On the contrary, it is theory which suggests what experiments should be performed. Furthermore, a theory may be generally accepted long before it can be confirmed by observational data. Copernicus advanced his heliocentric hypothesis three centuries before the rotation of the earth could be experimentally confirmed. And yet by the time that Foucault performed the decisive experiment around the middle of the nineteenth century, the only doubters who remained were the crackpots.

Similarly, Einstein's general theory of relativity was accepted long before convincing experimental confirmation was possible. Although experiments on the bending of light by the sun were performed as early as 1919, it wasn't until the 1960s and the 1970s that really accurate tests of the theory became possible. But most theoretical physicists had convinced themselves that the theory was correct long before these tests were made. Over and over again, physicists expressed the conviction that general relativity was "too beautiful to be false."

It is theoretical insight, not experimental research, that performs the greater part of the labor of "dismantling the universe." It is theory which demonstrates the falsity of naive commonsense ideas, and which perceives the hidden contradictions in scientific ideas that have been accepted for so long that they seem to be commonsense concepts also. If scientists did nothing but conduct experiments, our insights into the workings of nature could never have advanced beyond those of the medieval alchemists. It would be impossible to gain an understanding of the nature of the universe if scientists could not depend upon the insights that give birth to new theories. Admittedly, precise experiments must be performed if a theory is to be tested. However, theories themselves are products of the human imagination.

The scientist who constructs a new theory dismantles the universe. But of course, this is only part of his task. To dismantle it, he must find a way to replace it with something better. He must reassemble the universe in a new form.

Sometimes the process of reassembling the universe seems to be a never-ending task. As long as a theory is accepted, it is always possible to make further refinements in theoretical ideas, and to attempt to extend the range of the theory's application. It is still possible to do research in Newtonian mechanics, for example. One can even do theoretical research in mathematics by attempting to gain a better understanding of individual equations that arise out of Newton's theories.

Some scientific theories are so wide-ranging that they become the preoccupations of several generations of scientists. Darwin's theory of evolution is a good example. Although *Origin of Species* was published more than a hundred and twenty years ago, evolutionary biologists still engage in intensive research in an effort to understand the details of natural selection. Quantum mechanics must also be considered to be one of the most successful theories that science has known. Although the theory has existed since 1926, physicists are still laboring to understand all its implications.

Quantum mechanics has proved to be even more wide-ranging than its discoverers ever suspected. Today it is possible to use quantum mechanics not only to understand the behavior of atoms, but also to probe the nature of matter far below the atomic level. In 1926, physicists sought only to understand why atoms behaved as they did. Today, they commonly speak about the behavior of constituents of constituents of atoms, about neutrons, protons, mesons and quarks.

In its time, quantum mechanics was a revolutionary theory. In the fifty-odd years that have passed since it was discovered, it has fomented a whole series of revolutions. When Heisenberg, Schrödinger, Bohr and the other theoretical physicists of the 1920s argued about what quantum mechanics meant, they realized that they were tearing down long-held conceptions of physical reality. They did not suspect that they had started a continuing revolution. There was no way they could have known that each succeeding generation would reassemble science's conception of the subatomic world in a new and surprising way.

As is the case with most stories, the choice of a starting point is somewhat arbitrary. It would be possible to begin with the work of Heisenberg and Schrödinger or, alternatively, to start with the speculations of the ancient Greek philosophers concerning the ultimate nature of matter. But perhaps it would be best not to go to either extreme. A reasonable place to begin might be the year 1897.

Eighteen ninety-seven was the year in which the English physicist J. J. Thomson* discovered the electron, and showed that electrons had masses much smaller than that of hydrogen atoms. Since the hydrogen atom was the lightest known, it seemed only reasonable to suppose that the electron must be a constitutent of the atom. It was Thomson's discovery which provided the first hint that atoms were not indivisible, that the ultimate constituents of matter were something much smaller.

Since the electron was the only subatomic particle then known, Thomson made the natural assumption that atoms and electrons were the only particles that existed. He conceived the atom to be a sphere of positive electrical charge in which the negatively charged electrons were embedded. According to Thomson's theory, the electrons were free to move back and forth through the sphere. It was presumably changes in the vibrations of the electrons in an atom that were responsible for the emission of light.

But this commonsense conclusion led to difficulties. Physicists could not explain why matter emitted light as it did. If one assumed that vibrating electrons were responsible, one could not avoid the conclusion that infinite quantities of radiation would be emitted at short wavelengths.

This was the problem that so troubled Planck. Although Planck managed to remove the difficulty by assuming that light must be emitted in bundles of energy, or "quanta," he gained no insight into the problem of what atoms were like. There was something wrong with Thomson's theory, but no one knew what the defect was.

Part of the problem was solved when Rutherford discovered the atomic nucleus in 1911. But new difficulties arose in its place. If one assumed that an atom was made up of a nucleus that was surrounded by orbiting electrons—and it seemed obvious that Rutherford's experiments could imply nothing else—it was still

* His full name was Joseph John Thomson. However, physicists always refer to him as "J.J."

not apparent why light energy should be emitted in quanta.

The problem was finally solved when quantum mechanics was formulated in 1926. Yet the solution to this particular problem was only the beginning. The application of quantum mechanics to the structure of atoms was to lead to some of the most astonishing discoveries of modern physics.

The discoveries came about because every time that quantum mechanics answered a question, new ones arose in its place. No sooner had physicists found a satisfactory explanation for the behavior of atoms than they began to realize that it was not so easy to tell just what atoms were made of.

One of the first questions to arise was that of the nature of the atomic nucleus. Now, the nucleus of hydrogen, the simplest atom, presented no problems. Since atoms were electrically neutral, it was only reasonable to assume that hydrogen's single electron must be orbiting a positively charged particle. This particle was called the *proton*.

The case of the next atom, helium, was somewhat more complicated. Helium atoms were approximately four times as heavy as those of hydrogen. Since the mass of the electron was small (a proton is 1,836 times as heavy), the extra mass had to be concentrated in the nucleus. This seemed to imply that a helium nucleus contained four protons. However, it was known that the helium nucleus had a positive charge of two units, not four. Therefore it was assumed that the helium nucleus must contain four protons and two electrons. The negative electrical charges of the electrons would cancel out two of the positive charges, giving a net charge of plus two. The two electrons would not add significant mass to the nucleus, so it would still be approximately four times as heavy as that of hydrogen.

There was only one problem with this theory. Quantum mechanics implied that if a light particle like an electron was confined within a nucleus, it would bounce back and forth at a very high velocity. Its energy of motion would be so great that it would quickly bounce right out of the nucleus.

Did the nucleus contain electrons or didn't it? It was not possible to answer this question by determining the helium nucleus' mass. The measurement could be carried out accurately enough; that was no problem. The difficulty was that according to Einstein's equation $E = mc^2$, the mass of a nucleus was not the sum of the masses of its constituent particles. If special relativity was correct, the energy that bound the particles in a nucleus together had mass too. But at the time, nothing was known about the forces that glued the protons in a nucleus together. Hence, it was impossible to calculate this binding energy, or to determine its mass equivalent.

The problem was solved in 1932 when the English physicist James Chadwick discovered a third particle, the neutron. The neutron had a mass which was nearly the same as that of the proton. The most significant difference between the two particles was that the neutron did not have any electrical charge.

After the neutron was discovered, it was quickly realized that it was not necessary to assume that electrons existed within the nucleus after all. The helium nucleus could be made up of two protons and two neutrons, rather than four protons and two electrons. It appeared that the problem of the structure of matter had been solved. The three particles—neutron, proton and electron— seemed to account for everything. The physicists of the day confidently made the assumption that all atomic nuclei were composed of neutrons and protons. When electrons were added, one had atoms. There was nothing else.

The physicists who constructed this picture of matter had no way of knowing how wrong they were. The proton, the neutron and the electron were only the first in what was eventually to become a seemingly endless list of subatomic particles. No sooner had physicists constructed a reasonably parsimonious model of the structure of matter than they discovered that another subatomic particle existed, and another, and yet another.

The proliferation began in 1932, the same year that the neutron was discovered. In that year, the American physicist Carl Anderson discovered the *positron*, a particle which had the same mass as the electron, but which had a positive, rather than a negative, electrical charge.

The existence of the positron had been predicted, on theoretical grounds, in 1930 by the English physicist P. A. M. (Paul Adrien Maurice) Dirac. In 1928, Dirac had found a way to combine quantum mechanics with the special theory of relativity. Two years later, in 1930, he pointed out that his new, composite theory implied that there should exist an odd kind of electron which carried a positive charge. According to Dirac, the addition of Einstein's $E = mc^2$ to the formalism of quantum mechanics led to a theory which predicted that a pair of particles—an electron and a positron—could be created out of pure energy. Furthermore, the particles should be able to annihilate each other, yielding the original energy again. Finally, a particle that was created at the same time as an electron had to have a positive charge. Although matter was created out of energy, there was no reason to think that electrical charge could be also. In fact, observations indicated that it could not. Matter was ordinarily electrically neutral, indicating that there existed no natural processes which could cause an excess of one kind of charge or the other. Whenever one did give a positive charge to an object, an equal negative charge was always imparted to something else. Dirac assumed that the same was true of the creation of matter from energy. If positively and negatively charged particles were always created together, the net amount of charge created would be zero. It was simply a matter of adding +1 and −1 together.

The positron is said to be the *antiparticle* of the electron. Today it is known that every particle has an antiparticle. There are antiprotons and antineutrons, antimesons and antiquarks. Theoretically, it should be possible to make *antimatter* out of antiparticles. Antimatter is nothing more than matter in which electrons are replaced by positrons, protons by antiprotons and so on. Antimatter could not exist on the surface of the earth, because matter and antimatter would annihilate each other completely whenever they came into contact. In fact, a matter-antimatter explosion would be much more powerful than that of a hydrogen bomb. In a fusion bomb explosion, only a small fraction of the matter present is changed into energy. When matter and antimatter annihilate each other, on the other hand, all the mass that is present is converted.

It is conceivable that entire galaxies could be made out of antimatter. However, it is not very likely. Galaxies sometimes collide, and if antimatter galaxies existed, it is only reasonable to suppose that they would sometimes collide with galaxies made out of ordinary matter. If this happened, enormous amounts of energy would be released. Few, if any, stars would annihilate each other. The distances between stars are so great that there is little chance of their crashing into one another when galaxies collide. However, the interstellar gas contained in the matter and antimatter galaxies would intermix. This would cause annihilation to take place on a large scale. Since such annihilation has never been observed, most astronomers conclude that antimatter galaxies do not exist. For some reason, our universe seems to be made up wholly or primarily of matter.

Dirac's theory of electrons and positrons was not the only important development that took place in the field of quantum mechanics in the period between 1926 and 1930. During the late 1920s, physicists began to develop a theory called *quantum electrodynamics*. Quantum electrodynamics, or QED as it is often called, was not the work of any individual physicist. During the early years of its development, Dirac, Heisenberg and Pauli all made contributions. In later years, a number of other physicists were to develop the theory still further.

It was quantum electrodynamics that introduced the concept of the virtual particle. According to QED, it is the exchange of virtual particles that is responsible for subatomic forces. For example, the electrical repulsion between two electrons comes about because one electron will emit a virtual photon that is absorbed by the other. The exchange of photons nudges the two particles apart. The attractive force between electrons and protons is explained in the same manner. In this case, the exchange of virtual photons pushes the negatively charged electron and the positively charged proton together.

If quantum electrodynamics is correct, all charged particles are emitting and absorbing virtual photons at every moment. Sometimes these photons are absorbed by the particle that emits them, and sometimes they interact with other charged particles that are nearby. When the theory is worked out in detail, it is seen that the resulting forces can be either attractive or repulsive, depending upon the charges of the particles that interact with them.

There is an analogy which can be used to explain why the exchange of photons should produce a force that is sometimes attractive and sometimes repulsive. It goes something like this: Imagine that two people are standing on a frictionless surface, or on one that is nearly frictionless, such as a sheet of very slick ice. Imagine also that they throw a football to each other. It is obvious that the forces exerted when the football is thrown or caught will push them farther and farther apart. This would correspond to the repulsion experienced by two like (i.e., two positive or two negative) charges. Now suppose that the same two people turn their backs to each other and begin playing catch with a boomerang. Since a boomerang can curve around and change direction, tossing it back and forth will cause the two individuals to slide toward each other.

There is no fundamental difference between the virtual photons that cause charged particles to interact with each other and the photons which make up light and other forms of electromagnetic radiation. Virtual photons exist for so short a period of time that they cannot be observed directly, but in all other respects they are the same as the real ones. There is, by the way, no doubt about the virtual photons' existence. Quantum electrodynamics makes numerical predictions about the behavior of matter that can be verified in the laboratory to a high degree of accuracy.

Quantum electrodynamics is a very attractive theory. It is extremely parsimonious: it succeeds in explaining a wide variety of phenomena with a small number of assumptions. It accounts for the interaction of charged particles, certain details of the emission of light from atoms, and various magnetic phenomena. All of these phenomena are manifestations of a fundamental force of nature called the *electromagnetic interaction.* The electromagnetic interaction governs the emission and absorption of both real and virtual photons. It is responsible for the photons of light that can travel through space for billions of years and as well as for the virtual photons that exist for only a tiny fraction of a second.

The electromagnetic interaction is only one of the forces observed in nature. The gravitational interaction is another. But the list does not end here. It is obvious that there must be at least one other force, and possibly more. After all, neither the electromagnetic nor the gravitational force could possibly bind neutrons and protons together in a nucleus. Gravity is much too weak to accomplish that. The electromagnetic force, on the other hand, would cause protons to repel each other, and would not act on the electrically neutral neutrons at all. If the electromagnetic and gravitational forces were the only ones that existed, there could be no atoms other than those of hydrogen, and the universe would contain no stars, no planets and no life.

So physicists knew from the beginning that there must be a nuclear force. However, they had no idea what this force was like. All they knew was that there was something which caused the particles in the nucleus to stick together.

In 1935, the Japanese physicist Hideki Yukawa pointed out that there must exist a new, as-yet-undiscovered particle that was associated with the nuclear force. Working from analogy with quantum electrodynamics, he suggested that neutrons and protons bound themselves to each other by emitting and absorbing virtual particles. Yukawa calculated that these particles should have a mass that was roughly 200 times that of an electron (or about one-ninth that of a neutron or proton). This hypothetical particle was named the *meson.*

There was every reason to believe that mesons could be detected. Like virtual photons, virtual mesons lived for too short a time to be seen. However, there was no reason why real mesons could not be produced in certain types of nuclear reactions. After all, real photons existed.

In the same year that Yukawa published his theory, Carl Anderson discovered a particle that seemed to be Yukawa's meson. The particle, which was discovered in cosmic rays, seemed to have approximately the right mass. At first, this particle was called the *mu meson* (*mu* is one of the letters of the Greek alphabet). Today it is referred to as the *muon*. The "meson" has been dropped because shortly after Anderson made his discovery, physicists realized that the muon was not Yukawa's meson at all. On the contrary, it was a particle whose properties were very much like those of an electron. In fact, the only difference between the muon and the electron seemed to be that the former had a much larger mass.

Yukawa's particle was finally discovered in 1947. It is nothing other than the pion, which was discussed briefly in Chapter 7. Since the pion is the meson that Yukawa postulated, *pi meson* is still used as an alternative name.

If the proton, neutron, electron, muon, photon and pion, and their antiparticles, were the only ones that existed, physicists could have lived with the fact. To be sure, things did not seem as simple as they had when it appeared that everything was made up of just two or three elementary particles. However, six was really not so bad. And the existence of antiparticles really didn't complicate things all that much. Some particles, such as the photon, were their own antiparticles. So the number of fundamental entities was not quite doubled. In any case, antiparticles did not seem to be anything more than a kind of mirror image of those which made up ordinary matter.

Unfortunately, matters rapidly became more complicated. Studies of cosmic-ray showers (cosmic rays are not really "rays," but rather showers of particles) turned up new kinds of mesons. Beginning in 1948, additional particles began to be produced in the laboratory by cyclotrons. Physicists soon discovered that there were numerous different kinds of mesons, and a large number of different varieties of *baryons* (a baryon is a particle which bears a resemblance to the neutron and proton) as well.

In 1953, the *neutrino* was detected. This particle seemed to be the strangest of them all. Postulated by Wolfgang Pauli in 1930, the neutrino was a particle that had no charge and no mass. It traveled at the speed of light,* and it rarely interacted with matter. The neutrino was a participant in a certain type of radioactive decay. Its only function seemed to be to carry off part of the energy that was produced. If neutrinos existed, it followed that there also had to be *antineutrinos*. The only difference between neutrinos and antineutrinos seemed to be that they had opposite spin. One always spun in a clockwise direction as it flew away from the particle or nucleus that emitted it; the other always spun in the opposite manner.

By the end of the 1950s, physicists had discovered so many different particles that it was obvious that the term *elementary particle* was a misnomer. There were hundreds of different kinds of mesons, and hundreds of different kinds of baryons. Since it was reasonable to assume that yet more particles would be discovered in the future, it was possible that these "elementary" constituents of matter might even be infinite in number.

The situation was not unlike that encountered by the astronomers of the sixteenth and seventeenth centuries. Like the Ptolemaic system of astronomy, a theory of matter which depended upon the existence of countless fundamental particles was just too complicated to be reasonable. Physicists were dismayed by the complexity they had uncovered; they found it hard to believe that nature could really be like that at its lowest level.

To give some order to the growing chaos, particles were classified according to the forces they experienced. By this time, it had been known for quite some while that the interactions between known particles depended upon four fundamental forces: the

---

* Massless particles can travel at the speed of light. One could go so far as to say that they must. A zero-mass particle that did not travel at light velocity would have zero energy; it could not even be said to exist. Neutrinos, by the way, are not unique in this respect. Photons are massless also. They travel at the speed of light for the simple reason that they *are* light.

Recently, it has been suggested that there might be reasons for supposing that neutrinos have masses that are very small, but not zero. If this turns out to be true, then we will have to conclude that although they travel very fast, they do not quite attain light velocity. A neutrino with mass would be subject to the same limitations as any massive body.

gravitational interaction, the electromagnetic interaction and the *strong* and *weak* nuclear interactions.

The strong interaction was the force between nuclear particles that was mediated by mesons. The weak interaction was a second nuclear force which was responsible for radioactive decay. The weak interaction, incidentally, is as necessary to our existence as the strong and electromagnetic forces. Without it, the nuclear reactions that take place in the sun could not continue. Our sun makes use of both the strong and the weak interactions when it "burns" its nuclear fuel to make light and heat.

It was observed that although most of the known particles felt the strong force, some did not. Those that did were called *hadrons*. The hadron family was made up of baryons and mesons. Particles that interacted via the weak force, but not the strong one, were called *leptons*. The lepton family was made up of the electron, the muon, the neutrino and their antiparticles when the classification was devised. Since that time, the *tau*, a particle similar to the electron and the muon, has been added, and it has become apparent that there are three different kinds of neutrino.

The photon, which felt neither the strong nor the weak force, was placed in a class by itself, as was the *graviton*, the hypothetical particle that is associated with the gravitational interaction.

The hadrons caused problems. If there were only six leptons, that was not so bad. It was possible to think of all six as "elementary." Although a smaller number would have been better, six was something that one could live with. The photon and the graviton presented no problem either; there was only one of each. Something, however, had to be done about the thousands of baryons and mesons. Physicists realized that they could not pretend to have a reasonable theory until they could explain why there should be so many.

The first step was to invent classification schemes that would allow baryons and mesons to be grouped in families. The most successful of these schemes, called the *eightfold way*, was independently discovered by the American physicist Murray Gell-Mann and the Israeli intelligence-officer-turned-physicist Yuval

Ne'eman in 1961. It was called the eightfold way because it grouped the most commonly observed baryons and mesons in sets of eight. The name was a kind of pun; the original eightfold way was a recipe for enlightenment devised by the Buddha in the fifth or sixth century B.C.

The next step was to try to find out why this particular classification scheme worked. If particles exhibited similarities that caused them to fit together in families, there had to be some underlying principle that would explain why the grouping seemed so natural. This principle was independently discovered by Gell-Mann and by the American physicist George Zweig in 1964. It was found that the eightfold way could be explained if one assumed that baryons and mesons were made up of particles that were still more fundamental.

Zweig wanted to call the new particles "aces." Gell-Mann named them *quarks*. It was the latter term that was adopted. The name *quark* is taken from a line in James Joyce's *Finnegans Wake*, "Three quarks for Muster Mark." The passage in Joyce's novel is a reference to the cuckolding of King Mark in the legend of Tristram and Iseult.

In the original version of the quark theory, there were three quarks, called *up, down* and *strange*. The name *strange* should not be taken literally. *Strangeness* is a whimsical name that has been given to a mathematical property possessed by certain particles; it was one of the properties that made the eightfold-way classification scheme possible.

If there were three quarks, there also had to be three *antiquarks*. These were called *antiup, antidown* and *antistrange*. According to the theory, these six particles made up all baryons and all mesons. Baryons were made of three quarks, while mesons were composed of a quark and an antiquark. For example, the proton contained one down and two up quarks, while the constituents of the positively charged pion were an up and an antidown.

According to the theory, quarks had electrical charges that were either one-third or two-thirds of those possessed by particles like the proton and electron. The up quark had an electrical charge of $+\frac{2}{3}$, while the down and strange quarks had charges of $-\frac{1}{3}$. When quarks combined, they had to do so in such a way that the resulting charge was a whole number. For example, the two up quarks in the proton contributed a charge of $+\frac{2}{3}$ each. This gave $+\frac{4}{3}$. When one subtracted the $-\frac{1}{3}$ charge on the down quark, the result was $\frac{3}{3}$, or 1. In the positive pion, the $+\frac{2}{3}$ charge of the up quark and the $+\frac{1}{3}$ charge of the antidown also added to give 1. In particles such as the neutron, which had no charge, the charges of the constituent quarks added up to zero. Yet other combinations of quarks could give charges of -1, -2 or $+2$.

At first, many physicists regarded quarks as nothing more than a useful mathematical fiction that could be used to impose order upon the chaotic world of subnuclear particles. They didn't believe that the quarks corresponded to anything real. When attempts to find free quarks in nature failed, their suspicions were confirmed. The quark, they concluded, was an imaginary particle that made the job of doing physics somewhat easier.

In 1968, however, it was found that quarks possessed a greater degree of reality than these physicists had suspected. In that year, an experiment was performed which indicated that protons, at least, seemed to be composite particles. Beams of high-energy electrons were directed at proton targets in order to probe the internal structure of the latter. It was found that the protons seemed to contain pointlike charges. The quark hypothesis had been confirmed.

For a while, it appeared that physicists had at last solved the riddle of the nature of matter. But the euphoria was not to last. Matters quickly began to get complicated again. A fourth quark was discovered in 1974, and a fifth in 1978. These were called the *charmed* quark and the *bottom* quark respectively. Furthermore, there were theoretical reasons for suspecting the existence of a sixth, called the *top** quark. And of course, it was possible that there might be yet others. Physicists know of no reasons why the list of quarks would have to come to an end at six.

---

* For a while, there was a movement to name top and bottom "truth" and "beauty"; however, these names didn't stick.

By the time the bottom quark had been discovered, the quark model had already acquired an additional complication. Theoretical investigations of the nature of the forces between quarks led to the conclusion that quarks had to come in three different *colors*. Here the term *color* has nothing to do with the color of ordinary objects. A lemon is yellow and an apple is red because their constituent molecules reflect light of certain wavelengths that are interpreted as "red" and "yellow" by the human eye and nervous system. Quarks, which are constituents of atoms, do not emit or reflect light. Thus they cannot have color in the ordinary sense of the word.

The concept of color, as it is applied to quarks, is analogous to that of electrical charge. Just as electrical attraction binds electrons to the nuclei of atoms, color charges bind quarks together in baryons and mesons. The major difference is that there are three color charges, *blue, green* and *red*, while electrical charges exist in only two forms, positive and negative.

Color charge is a property only of quarks, not of particles that are made up of quarks. Just as an atom is electrically neutral, baryons and mesons are colorless. In the proton, for example, the colors of the three quarks—one blue, one green and one red— cancel out in the same way that the three primary colors of light will combine to produce white. Although a meson contains only two quarks, color cancels here also. This is made possible by the fact that one of the constituents of a meson is an antiquark which possesses an *anticolor*. An antiquark that is colored *antired*, for example, will combine with a red quark to produce white. The idea of anticolor is not as bizarre as it sounds. It is analogous to the idea of complementary colors of light. Yellow and blue light, for example, combine to produce white.* The color yellow is said to be the *complement* of blue.

The net result of all this is that the original three quarks have become eighteen. Quarks come in six *flavors*: up, down, strange, charmed, bottom and top. A quark of one of these flavors can

---

* Of course, yellow and blue *pigments* combine to form green. However, the addition of colors of light, which is the more basic process, works in a slightly different way.

have any of three colors: red, blue and green. Since three times
six is eighteen, there are eighteen different possibilities. If antiquarks are counted separately, the number must be doubled to
thirty-six.

The theory which makes use of the concept of color charges to
explain the forces between quarks is called *quantum chromo-
dynamics,* or QCD. It is the analog of quantum electrodynamics,
the theory that was developed to explain the electromagnetic
interaction.

Since every force which exists in nature is associated with a
particle that carries the force, it follows that the color forces must
be mediated by some new kind of particle. Such particles are
called *gluons.* The term is given in the plural because QCD predicts that there should be not one gluon, but eight. This really
isn't very surprising. It is true that the electromagnetic interaction
is associated with only one kind of force-carrying particle. However, there are only two kinds of electric charge, while quarks
come in three colors, and have three anticolors as well.

The situation of particle physics has become nearly as chaotic
as it was in 1960. Physicists have sought an underlying simplicity
in nature, and they have found themselves confronted with a
Ptolemaic complexity instead. The number of known particles has
again become unreasonably large. Depending upon whether one
counts antiparticles and different colors separately, there are six
or eighteen or thirty-six quarks. That is only a lower bound; there
could conceivably be more. Then there are eight gluons. Six different leptons are known, and of course they all have antiparticles
too. There are several different particles associated with the weak
interaction. Finally, we have the photon and the graviton. Although the graviton has not been detected, physicists do not
doubt that it exists. In addition, there is a possibility that there
might be a second particle, called the *gravitino,* that is also associated with gravity. Yet other particles are theoretical possibilities. One of the more intriguing of these is the *magnetic monopole,*
an isolated north or south magnetic pole.

Some possible simplifications have been pointed out. It has
been suggested, for example, that quarks might be made up of

yet more fundamental particles. However, this suggestion has not turned out to be very fruitful so far. The problem is that very large particle accelerators are required if one wants to obtain evidence for the existence of quarks, and even larger ones would be necessary if one wanted to probe the nature of matter at a yet lower level. The largest accelerators now measure approximately 2 kilometers in diameter. Since there are limitations on the amount of funding that governments are willing to provide, this is probably close to the practical limit. An accelerator large enough to probe the internal structure of quarks—if indeed they have any—would have to be of unreasonable size. An accelerator built around the circumference of the earth might not be powerful enough.

In all probability, the question of whether quarks have an internal structure will not be answered in the immediate future. If it is eventually discovered that they do, this might not be the end of the quest for an understanding of matter. It is possible that there might be a very large, or even infinite, number of levels of matter. There is no reason why quarks could not have more fundamental constituents, or why these constituents could not be made up of something still smaller, and so on. Probing the nature of matter is like peeling an onion layer by layer. We have no idea how many layers the onion might have.

As we have seen, the search for the fundamental constituents of matter has run into trouble twice—once around the end of the 1950s, and again during the 1970s. On both occasions, the problem that was encountered was the same. As soon as physicists had discovered the properties of the basic constituents of matter, they found that these supposedly "elementary" particles were multiplying beyond the bounds of reason. It seemed that every time they narrowed things down to just a few particles, those few immediately turned into dozens.

As a result, theoretical physics is currently trying a somewhat different approach. Attempts are being made to discover a single theory that will explain the four* fundamental forces of nature. If

* Since the strong force is nothing more than a manifestation of the color force between quarks, there are still only four.

this can be done (and if the forces don't suddenly begin to multiply the way particles have!), it might be possible to gain some clues as to why particle physics seems to be so complex. It might yet prove to be possible to discern a Copernican simplicity amid the complex interactions engaged in by particles that are produced in modern accelerators.

Einstein spent decades trying to find a unified field theory. In the end, he had to admit that he had failed. Furthermore, the problem that Einstein set himself is somewhat simpler than the one faced by physicists today. He was trying to unify just two forces—gravity and electromagnetism—rather than four. However, the problem is not as hopeless as one would think. There are difficulties, to be sure, but new theoretical methods have been developed, and the difficulties are gradually being overcome.

In fact, the first step has already been taken. In 1967, the American physicist Steven Weinberg and Abdus Salam, a Pakistani physicist teaching at the Imperial College in London, proposed a new theory which unified the weak and electromagnetic interactions. According to the Weinberg-Salam theory, the two interactions were nothing more than different aspects of a single fundamental force. The theory was confirmed in 1973 when evidence for a new particle, called Z, which had been predicted by the theory, was discovered. The theory also predicted that quarks should come in pairs. At the time, only the up, down and strange quarks were known. Up and down could be grouped together, but the strange quark did not seem to be paired with anything. When the charmed quark was discovered, it was considered to be an additional confirmation of the theory. It is the Weinberg-Salam theory, by the way, that has convinced physicists that there must be a top quark that is paired with the bottom quark, even though top has not yet been discovered.

The Weinberg-Salam theory reduced the four fundamental forces to three, which were now called the gravitational, the strong and the *electroweak*. Obviously, the next step would be to reduce these three to two.

So far, this task has not been accomplished. There currently exist a number of theories which attempt to unify the strong and electroweak interactions. But these *grand unified theories* (often referred to as GUTs) are quite speculative. They have not been confirmed experimentally, and no one knows which, if any, of them might be correct.

The grand unified theories do, however, make certain predictions which could help to solve long-standing problems in physics if they turn out to be true. For example, the GUTs seem to imply that the proton might not be as stable a particle as physicists had thought.

Ever since the proton was discovered, it had always been thought to be one of the few perfectly stable constituents of matter. A free neutron will decay into a proton, a positron and a neutrino in about fifteen minutes.* But a free proton will remain a proton, or so it was thought. It is true that protons can "trade places" with neutrons by emitting and absorbing mesons. But however many reactions of this sort take place, one always ends up with the same number of protons that were there at the beginning.

But according to the grand unified theories, there should be a process which can cause protons to decay (into a pion and a positron, for example). This process would be unlike the emission and absorption of mesons because it would cause protons to disappear.† The decay products would go their separate ways and participate in other reactions. Only on rare occasions would they form protons again.

If the proton does decay, it must have a very long lifetime. Otherwise we would observe matter disintegrating in the laboratory. Hydrogen atoms, for example, would suddenly disappear, and leptons and mesons would appear in their place.

The grand unified theories predict that protons should have a half-life of about 10^{32} years. Now, 10^{32} years is quite a long time. It is approximately 10 billion trillion times greater than the

* This is true only of free neutrons; it doesn't happen when they are bound in nuclei.

† The more technically minded reader may wish to know that the GUTs imply that the law of baryon conservation can be violated.

present age of the universe. One might think that it would be impossible to detect decays that take place so rarely. But this is not really the case. If one could assemble together 10^{32} protons, then it would be necessary to wait only one year before one of them decayed. And if one had 52×10^{32} protons, the waiting time would be reduced to one week. Now, it so happens that there are about 10^{32} protons in 300 tons of matter. Thus, the experiment is difficult but not impossible. If protons do decay, this fact may be experimentally confirmed at almost any time during the next several years.

One of the things that make the idea of proton decay so attractive is that it could explain why no antimatter has been observed in the universe. The reason that the antimatter problem is a puzzle is that until recently, physicists believed that protons and antiprotons had to be created in equal numbers. If the matter that exists today was created in the big bang, then matter and antimatter would exist in equal amounts.

But if the proton can decay into mesons and leptons, then the reverse process could presumably take place. If it can, then protons could have been created in this manner in the big-bang fireball. Naturally, proton-antiproton pairs would also be created by the more common process. But these would eventually annihilate each other, and only protons would remain in significant numbers. In other words, the GUTs seem to indicate that matter and antimatter do not always have to be created together. Sometimes matter can be created by itself.

If one or another of the grand unified theories is eventually confirmed, the next step will be to look for a theory that will unify all four forces, a *supergravity* theory. If such a theory is found, it might very well answer the question "Why are there so many particles?" A supergravity theory, in other words, might suggest that the observed particles are manifestations of some more fundamental principle.

Some attempts have been made to construct supergravity theories. So far, they have not been very successful. The theories that have been proposed do not seem to be able to account for all the particles that have been seen in experiments. In particular, supergravity theories cannot account for all the leptons and quarks.

Naturally, attempts have been made to remedy this. But no one has been able to come up with a theory that is very convincing. As a result, physicists have been trying out new assumptions, some of which are quite bizarre.

It has been shown, for example, that some of the problems associated with supergravity might be solved if one makes the assumption that space-time has eleven dimensions rather than four. The extra seven dimensions presumably cannot be observed because they are "curled up" to dimensions billionths of billionths the size of those of an atomic nucleus.

Extra, curled-up dimensions are really not so hard to imagine. If a two-dimensional sheet of paper is rolled up very tightly, it begins to resemble a one-dimensional rod. However, the fact that we can visualize curled-up dimensions does not imply that the concept is plausible.

The idea of seven extra dimensions sounds too much like an *ad hoc* assumption. If the only reason for assuming the extra dimensions is to make a theory work, then there are no good reasons for taking the hypothesis seriously. Fixing up supergravity theory with extra dimensions is too much like the activity that astronomers engaged in five hundred years ago, when they fixed up the Ptolemaic theory by adding extra epicycles and equants.

If theoretical physicists eventually discover some more compelling reason for assuming extra dimensions, then we will have to accept that they may exist. Possibly, work on supergravity will demonstrate that some even more bizarre assumption is necessary. If that happens, we may very well find that our commonsense conception of reality has been shattered once again.

In fact, we should probably expect that theoretical work on the unification of the forces will dismantle our conceptual universe in one way or another. This, as we have seen, has happened every time a major scientific advance has been made. And of course, if the universe is dismantled once again, scientists will find themselves busily engaged in the task of reassembling a new one.

EPILOGUE
HOW REAL IS "REAL"?

PHYSICISTS OFTEN SPEAK of "seeing" the subatomic particles that are produced in particle accelerators. The particles can be made to interact within a device called a *bubble chamber*, which is a large tank that has been filled with liquid hydrogen. As the particles pass through the hydrogen, they produce tiny bubbles. After these bubbles are photographed, the particle interactions that took place can be studied at leisure.

But are the particles really *seen*? Of course not. The only things that are observed are hydrogen bubbles that are strung together like beads. All physicists actually see is that some of the hydrogen in the tank has been converted into bubbles of vapor. The particles themselves are too small to be observed directly, even by the most powerful electron microscopes.

Protons and neutrons, leptons, photons and quarks don't "really exist" in the way that tables and chairs do. Their existence is inferred; physicists believe in them because they can be used to explain so many things that are observed in the laboratory. The subatomic particles are "real" in the sense that they are concepts which can be used to codify an enormous amount of data about natural phenomena. The theories which describe the behavior of particles provide us with *models* of reality.

If a scientific model is a successful one, it will give more or less accurate predictions when the right mathematical calculations are carried out. Scientists observe certain phenomena, such as the emission of light, and invent concepts that will explain them. They perform mathematical manipulations upon these concepts in order to find out what other phenomena can be predicted. But however successful their models are, they are never perfectly accurate. Each generation discovers that it must find a way to improve upon the model used by the previous one. The concept of indivisible atoms had to be replaced by a picture in which atoms were composed of electrons, protons and neutrons. The concept of a nucleon as an elementary particle was replaced by that of gluons and quarks.

Many scientists and philosophers of science regard the concepts of physics as maps or pictures rather than as real objects. This is what Bohr meant when he said that the quantum world did not exist. In Bohr's view, the only things that were "real" were the readings on pointers and dials in the laboratory. Electrons, probability waves and other subatomic entities were nothing but useful constructs.

If we regard scientific discovery as the construction of useful models, this would explain why theory has been more important than experiment so many times in the history of science. By itself, experiment tells us nothing. Theories are not constructed from generalizations derived from experimental data. On the contrary, they are products of human imagination. A theory is discovered when a scientist has an insight which shows him that it is possible to invent a model which will make it possible for data to be interpreted. Theory organizes data, and brings them together in a picture that human beings find comprehensible.

Scientific models are not entirely arbitrary. A theory must be able to codify data in a useful form, or it is useless. If one imagined that a flat earth was supported by four huge elephants standing on the back of an enormous turtle, he would have created a model. But this particular model could not be used to bring together data derived from observations of the earth's moon and the planets, or even observations of the ocean tides. A scientist who is constructing a theory cannot ignore the fact that there are data to be explained. He is no more free to imagine anything he likes than a poet is free to write a sonnet that has fifteen lines.

In the time of Kepler, Newton and Galileo, scientists believed that they were discovering exact "laws of nature." Today we realize that the discovery of exact laws is not possible. At best, a scientific theory is a set of approximations. The approximations, however, can be extremely accurate, and models can be developed that tie together an astonishing range of phenomena. Perhaps science does not really discover order in nature. Order may be something that is imposed upon nature by human minds. Nevertheless, this order can be an impressive one.

If the "order of the universe" is something that we invent, then it is somewhat easier to see why it is so obvious that a theory that is "just too unreasonable to be true" can so readily be discarded. Even when there exists no experimental evidence that would allow us to decide between two competing theories, it can still be obvious that one is a better model than the other.

For example, it would be possible to argue that the Ptolemaic system is really just as valid as the Copernican. If Einstein was correct, and the viewpoint of one observer is just as good as that of another, then there is no way we can prove that the sun does not really revolve around the earth. After all, from the standpoint of an earthbound observer, the heavens do rotate, and the sun does rise and set.

The only reason that scientists do not make this assumption is that it would lead to a theory that was anything but clear or logical or elegant. Not only would one have to deal with all the complications that Ptolemy encountered, one would also have to think up an explanation for the centrifugal force that causes the earth to bulge outward at the equator, and explain why terrestrial gravity is a little stronger at the poles. One would have to explain why the orbital velocities of distant stars appear to be so much greater than the speed of light, and it would be necessary to explain the origin of the *coriolis force* that imparts an eastward or westward drift to a missile that is fired over the North Pole.*

There is no doubt that it would be possible to explain all these things in one way or another. However, the resulting theory would be ugly; it would be anything but parsimonious. It would be as complicated as a theory designed to prove that the earth was really flat.

I often have the feeling that scientific discovery and the arts are more alike than we tend to think. As I pointed out in a previous chapter, it is not easy to demonstrate that there are any real connections. Nevertheless, I sometimes wonder if science could not be described as an attempt to impose order upon physical reality, which would make it rather similar to art, which could be described as an activity that imposes order upon all the myriad aspects of human experience.

Sometimes artistic trends and scientific theories exhibit similarities one would not expect. Poets too can dismantle the universe and recombine it in startling ways. For example, during most of Western history, mountains were regarded as fearsome objects that were the likely abodes of demonic creatures. But when the Romantic artists and poets began to see a kind of grandeur in mountainous landscapes, attitudes changed. When the Romantics showed their contemporaries that mountains could be experienced in a different way, men even began to think of climbing them. Mountaineering was, in fact, an invention of the Romantic age. As far as we know, no one had ever considered the idea previously.

Most works of art do not change our attitudes about physical objects. After all, art attempts to structure the universe of human experience, and emotional reactions to mountains and landscapes constitute only a small part of it. Nevertheless, the analogy between the Romantic restructuring of a certain part of the world and the Einsteinian restructuring of the universe is striking.

*At least these things would be a problem in the simple Newtonian approximation that astronomers use most of the time. They would naturally pose no problems in general relativity, where motion is relative.

When Einstein propounded his two theories of relativity, the universe also became a different kind of place. The general theory, in particular, made it possible to give up the idea of gravitational "action at a distance" that had so bothered the contemporaries of Newton. This mysterious "action" was replaced by the concept of changes in the geometry of space.

There are failed works of art which are grotesque rather than moving. Similarly, there are crackpot theories, and theories which are too complicated to be reasonable. Not all attempts to dismantle or restructure the universe are successful. Of those which are successful, not all are equally imposing. We admire the theories of Einstein because they restructure our conception of the universe in clear and logical ways, just as we admire the works of Shakespeare because so much of the universe of human experience is given form in his plays.

Perhaps this is a digression. But it is not an irrelevant one, I think. The events which take place on the magic isle in *The Tempest* have something of the character of a dream. There is a sense in which mesons and quarks and gluons are dreams too. Do they "really" exist? They almost certainly do not.

And yet one could say that they do exist in some sense of the word. They are real in the sense that they provide us with a map that allows us to find our way around in the chaotic world of subatomic phenomena. They are real in the sense that they give us a picture of this world that is astonishingly vivid and fruitful.

Models are useful because they show us that hidden connections exist in the universe around us. And after all, this is what science is all about. Science seeks to create pictures of the order in nature which are so logically elegant that we cannot doubt that they are true.

BIBLIOGRAPHY

Abell, George O., and Singer, Barry, eds. *Science and the Paranormal*. New York: Scribner, 1981.

Bernstein, Jeremy. *Einstein*. Harmondsworth, England: Penguin, 1976.

———. *Experiencing Science*. New York: Basic Books, 1978.

———. *Science Observed*. New York: Basic Books, 1982.

Beveridge, W. I. B. *The Art of Scientific Investigation*. New York: Vintage, n.d.

———. *Seeds of Discovery*. New York: Norton, 1980.

Bohm, David. *Wholeness and the Implicate Order*. London: Routledge & Kegan Paul, 1980.

Bohr, Niels. *Atomic Theory and the Description of Nature*. New York: Macmillan, 1934.

Brockman, John. *Afterwords*. Garden City, N.Y.: Anchor, 1973.

Bronowski, J. *The Common Sense of Science*. Cambridge, Mass.: Harvard University Press, 1978.

———. *The Identity of Man*. Garden City, N.Y.: Natural History Press, 1972.

———. *The Origins of Knowledge and Imagination*. New Haven: Yale University Press, 1978.

———. *Science and Human Values*. New York: Perennial Library, 1972.

———. *The Visionary Eye*. Cambridge, Mass.: MIT Press, 1978.

Burbidge, Geoffrey. "Do We Really Know How to Measure the Distance of Extragalactic Objects?" *Annals of the New York Academy of Sciences,* Vol. 375 (1981), pp. 123–56.

Capra, Fritjof. *The Tao of Physics.* New York: Bantam, 1977.

Clark, Ronald W. *Einstein.* New York: Avon, 1972.

Cline, Barbara Lovett. *The Questioners.* New York: Crowell, 1965.

Corliss, William R., ed. *Mysterious Universe.* Glenarm, Md.: The Sourcebook Project, 1979.

Cropper, William H. *The Quantum Physicists.* New York: Oxford University Press, 1970.

d'Abro, A. *The Rise of the New Physics.* 2 vols. New York: Dover, 1952.

Davies, P. C. W. *The Forces of Nature.* Cambridge, England: Cambridge University Press, 1979.

———. *The Physics of Time Asymmetry.* Berkeley: University of California Press, 1977.

———. *The Runaway Universe.* New York: Harper & Row, 1978.

———. *Space and Time in the Modern Universe.* Cambridge, England: Cambridge University Press, 1977.

Davis, Philip J., and Hersh, Reuben. *The Mathematical Experience.* Boston: Houghton Mifflin, 1981.

"The Debate Goes On." *The Sciences,* Vol. 22, No. 4 (April 1982), pp. 2–3.

d'Espagnet, Bernard. *Conceptual Foundations of Quantum Mechanics,* 2nd ed. Reading, Mass.: Benjamin, 1976.

———. "The Quantum Theory and Reality." *Scientific American,* Vol. 241, No. 5 (November 1979), pp. 158–81.

DeWitt, Bryce S., and Graham, Neill. *The Many-Worlds Interpretation of Quantum Mechanics.* Princeton: Princeton University Press, 1973.

Dossey, Larry. *Space, Time & Medicine.* Boulder, Colo.: Shambhala, 1982.

Dott, Robert H., Jr., and Batten, Roger L. *The Evolution of the Earth.* New York: McGraw-Hill, 1981.

Durrell, Lawrence. *A Key to Modern British Poetry.* Norman: University of Oklahoma Press, 1952.

Dyson, Freeman. *Disturbing the Universe.* New York: Harper & Row, 1979.

Eddington, Arthur. *The Philosophy of Physical Science*. Ann Arbor: University of Michigan Press, 1958.

Edmonds, M. G. "Quasars Resolved?" *Nature*, Vol. 295 (1982), p. 556.

Einstein, Albert, and Infeld, Leopold. *The Evolution of Physics*. New York: Simon and Schuster, 1966.

Einstein, Albert, et al. *The Principle of Relativity*. New York: Dover, 1952.

Feinberg, Gerald. *What Is the World Made Of?* Garden City, N.Y.: Anchor, 1978.

Ferris, Timothy. *The Red Limit*. New York: Morrow, 1977.

Feynman, Richard. *The Character of Physical Law*. Cambridge, Mass.: MIT Press, 1967.

Fleck, Ludwik. *Genesis and Development of a Scientific Fact*. Chicago: University of Chicago Press, 1979.

Franks, Felix. *Polywater*. Cambridge, Mass.: MIT Press, 1981.

French, A. P., ed. *Einstein: A Centenary Volume*. Cambridge, Mass.: Harvard University Press, 1979.

Galileo. *Dialogue Concerning the Two Chief World Systems*. Berkeley: University of California Press, 1967.

Gamow, George. *Thirty Years That Shook Physics*. Garden City, N.Y.: Anchor, 1966.

Gardner, Martin. *Fads and Fallacies*. New York: Dover, 1957.

———. *Science: Good, Bad and Bogus*. Buffalo, N.Y.: Prometheus, 1981.

Giedion, Sigfried. *Space, Time and Architecture*, 4th ed. Cambridge, Mass.: Harvard University Press, 1962.

Golden, Frederic. *The Moving Continents*. New York: Scribner, 1972.

Gregory, Richard L. *Mind in Science*. Cambridge, England: Cambridge University Press, 1981.

Gribbin, John. *Timewarps*. New York: Delacorte, 1979.

Grim, Patrick, ed. *Philosophy of Science and the Occult*. Albany: State University of New York Press, 1982.

Guillemin, Victor. *The Story of Quantum Mechanics*. New York: Scribner, 1968.

Hadamard, Jacques. *The Psychology of Invention in the Mathematical Field*. New York: Dover, 1954.

Hansel, C. E. M. *ESP and Parapsychology.* Buffalo, N.Y.: Prometheus, 1980.

Hardy, G. H. *A Mathematician's Apology.* London: Cambridge University Press, 1967.

Harrison, Edward R. *Cosmology.* Cambridge, England: Cambridge University Press, 1981.

Hawking, S. W., and Israel, W., eds. *General Relativity.* Cambridge, England: Cambridge University Press, 1979.

Heisenberg, Werner. *Physics and Beyond.* New York: Harper & Row, 1972.

———. *Physics and Philosophy.* New York: Harper & Row, 1958.

Hermann, Armin. *The Genesis of Quantum Theory (1899–1913).* Cambridge, Mass.: MIT Press, 1971.

Hoffman, Banesh. *Albert Einstein.* New York: New American Library, 1973.

Hofland, David. "The Great Redshift Debate." *The Sciences,* Vol. 21, No. 8 (October 1982), pp. 10–14.

Holton, Gerald. *Thematic Origins of Scientific Thought.* Cambridge, Mass.: Harvard University Press, 1975.

Hook, Sidney, ed. *Determinism and Freedom in the Age of Modern Science.* New York: New York University Press, 1958.

Huxley, Aldous. *Literature and Science.* New York: Harper & Row, 1963.

Jammer, Max. *The Conceptual Development of Quantum Mechanics.* New York: McGraw-Hill, 1966.

———. *The Philosophy of Quantum Mechanics.* New York: Wiley, 1974.

Kaufmann, William J., III. *The Cosmic Frontiers of General Relativity.* Boston: Little, Brown, 1977.

Kline, Morris. *Mathematics: The Loss of Certainty.* New York: Oxford University Press, 1980.

Klotz, Irving M. "The N-Ray Affair." *Scientific American,* Vol. 242, No. 5 (May 1980), pp. 168–75.

Koestler, Arthur. *The Act of Creation.* New York: Dell, 1967.

———. *The Sleepwalkers.* New York: Macmillan, 1959.

Kopel, Zdeněk. *The Realm of the Terrestrial Planets.* New York: Wiley, 1979.

Kuhn, Thomas S. *The Copernican Revolution.* Cambridge, Mass.: Harvard University Press, 1957.

————. *The Essential Tension.* Chicago: University of Chicago Press, 1977.

————. *The Structure of Scientific Revolutions.* Chicago: University of Chicago Press, 1970.

Lindsay, Robert Bruce. *The Nature of Physics.* Providence: Brown University Press, 1968.

Maffei, Paolo. *Monsters in the Sky.* Cambridge, Mass.: MIT Press, 1980.

Maran, Stephen P. "The Quasar Controversy Continues." *Natural History,* Vol. 91, No. 1 (January 1982), pp. 85–87.

Margenau, Henry. *The Nature of Physical Reality.* New York: McGraw-Hill, 1950.

Millar, Ronald. *The Piltdown Men.* New York: St. Martin's, 1972.

Moore, Ruth. *Niels Bohr.* New York: Knopf, 1966.

Morris, Richard. *The End of the World.* Garden City, N.Y.: Anchor, 1980.

————. *The Fate of the Universe.* New York: Playboy Press, 1982.

————. *Light.* Indianapolis: Bobbs-Merrill, 1979.

Mulvey, J. H., ed. *The Nature of Matter.* Oxford: Oxford University Press, 1981.

Pagels, Heinz R. *The Cosmic Code.* New York: Simon and Schuster, 1982.

Particles and Fields: Readings from Scientific American. San Francisco: Freeman, 1980.

Petersen, Aage. *Quantum Physics and the Philosophical Tradition.* Cambridge, Mass.: MIT Press, 1968.

Polanyi, Michael. *Personal Knowledge.* New York: Harper & Row, 1964.

Popper, Karl R. *The Logic of Scientific Discovery.* New York: Harper & Row, 1968.

Robinson, Arthur L. "Quantum Mechanics Passes Another Test." *Science,* Vol. 217 (1982), pp. 435–36.

Russell, Bertrand. *A History of Western Philosophy.* New York: Simon and Schuster, 1945.

————. *Mysticism and Logic.* Garden City, N.Y.: Anchor, n.d.

————. *The Scientific Outlook.* New York: Norton, 1962.

Sachs, Mendel. *The Search for a Theory of Matter.* New York: McGraw-Hill, 1971.

Sagan, Carl. *Broca's Brain.* New York: Random House, 1979.

Scheibe, Erhard. *The Logical Analysis of Quantum Mechanics.* Oxford: Permagon, 1973.

Schlegel, Richard. *Superposition and Interaction.* Chicago: University of Chicago Press, 1980.

Schneer, Cecil J. *Mind and Matter.* New York: Grove, 1969.

Schrödinger, Erwin. *What Is Life?* and *Mind and Matter.* Cambridge, England: Cambridge University Press, 1967.

Scott, William T. *Erwin Schrödinger.* Amherst: University of Massachusetts Press, 1967.

Segré, Emilio. *From X-Rays to Quarks.* San Francisco: Freeman, 1980.

Shipman, Harry L. *Black Holes, Quasars and the Universe,* 2nd ed. Boston: Houghton Mifflin, 1980.

Silk, Joseph. *The Big Bang.* San Francisco: Freeman, 1980.

Stapp, Henry Pierce. "The Copenhagen Interpretation." *American Journal of Physics,* Vol. 60 (1972), pp. 1098–1116.

Stevens, Charles F. "The Neuron." *Scientific American,* Vol. 241, No. 3 (September 1979), pp. 54–65.

Sullivan, Walter. *Continents in Motion.* New York: McGraw-Hill, 1974.

Talbot, Michael. *Mysticism and the New Physics.* New York: Bantam, 1980.

Tarling, Don, and Tarling, Maureen. *Continental Drift.* Garden City, N.Y.: Doubleday, 1971.

Taylor, Gordon Rattray. *The Natural History of the Mind.* Harmondsworth, England: Penguin, 1981.

Ter Haar, D. *The Old Quantum Theory.* Oxford: Permagon Press, 1967.

Toulmin, Stephen, and Goodfield, June. *The Architecture of Matter.* Chicago: University of Chicago Press, 1982.

————. *The Discovery of Time.* Chicago: University of Chicago Press, 1965.

————. *The Fabric of the Heavens.* New York: Harper & Brothers, 1961.

Trefil, James S. *From Atoms to Quarks*. New York: Scribner, 1980.

Velikovsky, Immanuel. *Worlds in Collision*. New York: Delta, 1965.

Wegener, Alfred. *The Origin of Continents and Oceans*. New York: Dover, 1966.

Weinberg, Steven. *The First Three Minutes*. New York: Basic Books, 1977.

Weiner, J. S. *The Piltdown Forgery*. New York: Dover, 1980.

Wigner, Eugene P. *Symmetries and Reflections*. Cambridge, Mass.: MIT Press, 1970.

Wilber, Ken, ed. *The Holographic Paradigm*. Boulder, Colo.: Shambhala, 1982.

Woolf, Harry, ed. *Some Strangeness in the Proportion*. Reading, Mass.: Addison-Wesley, 1980.

Young, J. Z. *Programs of the Brain*. Oxford: Oxford University Press, 1978.

Zukav, Gary. *The Dancing Wu Li Masters*. New York: Morrow, 1979.

INDEX

experiments, (*continued*)
  repeatability, 50, 133
  thought, 13–15, 63–64
  with X rays, 130–31

Fedyakin, Nikolai, 134
Finnegans Wake (Joyce), 194
flat-earth theory, 174
fossil magnetism, 108–9
fossils:
  Piltdown, 121–24, 129
  similarities of, 104, 105–6, 107
Foucault, Jean Bernard Léon, 99,
  182
free will, 54–60, 61, 77

galaxies:
  antimatter, 188
  collision of, 188
  expansion of, 69, 70
  gravitational attraction between,
  153
  quasars and, 117–20
Galilean relativity, 19–20
Galileo, 20, 88, 94–98, 100, 174, 181
gamma rays, 176
  mini black holes and, 150–51
Gamow, George, 77, 175–77, 178
Gardner, Martin, 51
Gauss, Karl Friedrich, 165
Geller, Uri, 51
Gell-Mann, Murray, 42, 193–94
geometrical theory, 90–91, 100
geometry:
  Euclidean, 26–27, 90
  non-Euclidean, 27–28
geophysicists, continental drift
  attacked by, 106, 107
Germer, Lester, 34
Giedion, Sigfried, 167–68
gluons, 162, 197
Gondwanaland, 104
grand unified theories (GUTs), 200–201
gravitational lens, discovery of, 26
gravitinos, 197
gravitons, 193, 197
gravity, 71, 153, 190, 193
  in black holes, 13, 14, 25–26, 27–28, 48, 151
  inverse-square law of, 57, 64, 174
  in Newtonian physics, 57, 64, 96,
  98, 125, 139, 144, 171, 174, 177

quantum, 48, 57
in relativity theory, 25–28, 48, 49,
  57, 161
Greek astronomical theories, 81–85

hadrons, 193
half-life:
  of mini black holes, 150
  of protons, 201
  of radium 224, 147, 148
Hardy, Thomas, 171–72
Heisenberg, Werner, 35–36, 76, 188
  uncertainty principle of, 38–39,
  72–73
heliocentric solar system, 9, 16, 82,
  86–100, 114
  experimental verification of, 98–100, 182
  religious opposition to, 94, 95, 96–97, 98
helium atoms, 35, 36
  in big-bang theory, 175, 176, 177
  nucleus of, 185–86
Hess, Harry H., 110–14
Hofmannsthal, Hugo von, 171
Hooke, Robert, 98
Hubble, Edwin, 69–70, 116
hydrogen atoms, 35, 36, 74, 79, 201
  nucleus of, 185
  as source of helium, 176

imagination:
  reality vs., 49, 52–53, 136
  see also creativity
Institute for Parapsychology, 51
Institute of Surface Chemistry, 134
intuition:
  in scientific discovery, 64–65, 70–71, 72, 79–80, 121
  uses and pitfalls of, 80–101
isostasy, principle of, 105–6
isotopic-spin states, 161

Joyce, James, 158, 167, 194
Jupiter:
  moons of, 97
  orbit of, 89, 90
  sphere of, 83
  Venus ejected from, 137, 140–41

Kaufmann, Walter, 67, 80, 114
Keats, John, 179